WAVE DYNAMICS

WAVE DYNAMICS

Editors

Snehashish Chakraverty
National Institute of Technology Rourkela, India

Perumandla Karunakar
Anurag University, Hyderabad, India

World Scientific

NEW JERSEY · LONDON · SINGAPORE · BEIJING · SHANGHAI · HONG KONG · TAIPEI · CHENNAI · TOKYO

Published by

World Scientific Publishing Co. Pte. Ltd.
5 Toh Tuck Link, Singapore 596224
USA office: 27 Warren Street, Suite 401-402, Hackensack, NJ 07601
UK office: 57 Shelton Street, Covent Garden, London WC2H 9HE

Library of Congress Cataloging-in-Publication Data
Names: Chakraverty, Snehashish, editor. | Karunakar, Perumandla, editor.
Title: Wave dynamics / editors, Snehashish Chakraverty, National Institute of Technology, Rourkela, India, Perumandla Karunakar, Anurag University, Hyderabad, India.
Description: New Jersey : World Scientific, [2022] | Includes bibliographical references and index.
Identifiers: LCCN 2021047954 | ISBN 9789811245350 (hardcover) | ISBN 9789811245367 (ebook) | ISBN 9789811245374 (ebook other)
Subjects: LCSH: Wave mechanics. | Electrodynamics.
Classification: LCC QC174.2 .W375 2022 | DDC 531/.1133--dc23/eng/20211108
LC record available at https://lccn.loc.gov/2021047954

British Library Cataloguing-in-Publication Data
A catalogue record for this book is available from the British Library.

For any available supplementary material, please visit
https://www.worldscientific.com/worldscibooks/10.1142/12503#t=suppl

Desk Editors: Balasubramanian Shanmugam/Steven Patt

Typeset by Stallion Press
Email: enquiries@stallionpress.com

Printed in Singapore

https://doi.org/10.1142/9789811245367_fmatter

Preface

Various types of waves include water, sound, electromagnetic, seismic, and shock, and these waves need to be analyzed and understood for different practical applications. Recently, development of novel theories of wave dynamics and supported numerical procedures, as well as exploration of scientific and engineering applications for these waves, has emerged as a challenge. Moreover, these waves are modeled by corresponding equations of motion, which are in general governed by various mathematical equations. At times, it is also difficult to model the physics of the waves, and in those cases, experimental studies are important. In this regard, this book is an attempt to consider the waves in detail to understand the various physical and mathematical phenomena. To the best of our knowledge, books covering recent mathematical as well as experimental investigations are scarce.

This book consists of 13 chapters discussing various types of wave models. As such, Chapters 1–4 are devoted to ocean/water/shallow water waves. Seismic and plane wave models are discussed in Chapters 5 and 6. Chapters 7 and 8 include the analysis of shock and traveling waves. Uncertainty analysis of wave motion problems is encountered in Chapters 9–11. Finally, Chapters 12 and 13 are dedicated to fractional wave models.

As regards, Chapter 1 is written by G. Uma who provided a platform for better understanding of nonlinear wave–wave interactions and its sensitivity to water depths, directional distribution, and resolution. An attempt is made by Mayuresh Vaze and Naresh Krishna Vissa to quantify the characteristics of ocean surface waves during the passage of tropical cyclones in Chapter 2. Prashant Kumar, Vinita, and Rajni contributed Chapter 3, where one-dimensional (1-D) and two-dimensional (2-D) analytical and numerical solutions of Boussinesq Equations (BEs) are obtained utilizing the concept of shallow water waves over variable water depth. Chapter 4 is by T. K. Pal, D. Datta, and R. K. Bajpai who provided the numerical solution of the governing KdV equation for shallow water waves using a robust and efficient numerical scheme called differential quadrature method (DQM), which has potential applications in solving higher-order differential equations.

Sagarika Mukhopadhyay and Akash Kharita investigated the seismic wave equations in Chapter 5. Solution of wave equations for both homogeneous and heterogeneous media are derived, and from them, travel time equations along raypaths that obey Snell's law are obtained for simple velocity structures in this chapter. Abhishek Kumar Singh and Sayantan Guha presented the micro-mechanics model of PFRC and illustrated some of its advantages in Chapter 6. Analytical study of the impacts of normal/shear initial stresses and rotation on energies carried by different reflected/transmitted waves at the interface of two dissimilar PFRCs are discussed here.

Chapter 7 is contributed by Bikash Sahoo and Sradharam Swain. Here, recent developments of shock waves, shock wave solutions, and traveling wave solutions in viscoelastic generalization of Burger's equation are discussed. It also includes the shock wave solutions of differential equations arising due to flow of some viscoelastic fluids of differential type. Further investigations are carried out on wave propagation through a single resonator, periodically arranged resonators, and a multi-resonator unit structure by J. Dandsena and D. P. Jena in Chapter 8.

Chapter 9 is written by P. Karunakar and S. Chakraverty and they provided the solution of interval–modified Kawahara differential equation (mKDE) that describes the nonlinear water waves in long-wavelength regime using homotopy perturbation transform method (HPTM). The natural convection of a non-Newtonian fluid, sodium alginate, contaminated with copper nanoparticles between parallel plates with imprecisely defined parameters has been investigated by P. S. Sangeetha and Sukanta Nayak in Chapter 10, using interval homotopy perturbation method (IHPM). In Chapter 11, an interval-midpoint approach is used by T. D. Rao and S. Chakraverty for the mathematical modeling of uncertain radon transport mechanisms for predicting the anomalous behavior of the radon data generated in a soil chamber.

An analytical technique namely modified homotopy analysis transform method (MHATM) is used to solve time-fractional coupled shallow water equations (SWEs) that illustrate the flow below a pressure surface in a fluid in Chapter 12, which is contributed by Amit Kumar. Finally, Shweta Dubey and S. Chakraverty found approximate solutions in the form of convergent series for both 1-D and 2-D fractional wave equations using homotopy perturbation method (HPM) in Chapter 13.

This book is mainly written for undergraduates, graduates, researchers, industry, and faculties. The book covers various analytical and numerical/computational methods for solving different models governing water, sound, seismic, and shock waves. Further, uncertain and fractional models are also addressed to handle the wave equations in challenging areas which may make this book unique. The book may also be used for teaching classical and advanced wave analysis to graduates, undergraduates, and researchers.

Editors would like to thank all the authors who have contributed their chapters on time. Finally, the Editors do appreciate the whole team of World Scientific for their efforts, help, and support for the success of this challenging project and for publishing this important book as per schedule.

https://doi.org/10.1142/9789811245367_fmatter

About the Editors

S. Chakraverty has 30 years of experience as a researcher and teacher. Presently, he is working in the Department of Mathematics (Applied Mathematics Group), National Institute of Technology Rourkela, Odisha as a Senior (Higher Administrative Grade) Professor. Prior to this, he was with CSIR-Central Building Research Institute, Roorkee, India. After completing graduation from St. Columba's College (Ranchi University), his career started from the University of Roorkee (Now, Indian Institute of Technology Roorkee) and did M.Sc. (Mathematics) and M. Phil. (Computer Applications) from there securing First positions in the university. Dr. Chakraverty received his Ph.D. from IIT Roorkee in 1992. Thereafter, he did his post-doctoral research at the Institute of Sound and Vibration Research (ISVR), University of Southampton, U.K. and at the Faculty of Engineering and Computer Science, Concordia University, Canada. He was also a visiting professor at Concordia and McGill universities, Canada, during (1997–1999) and visiting professor of University of Johannesburg, South Africa during (2011–2014). He has authored/co-authored 23 books, published 386 research papers (till date) in journals and conferences and two books are ongoing. He is in the Editorial Boards of various international journals, book series, and conferences. Prof. Chakraverty is the Chief Editor of *International Journal of Fuzzy Computation and Modelling* (IJFCM), Inderscience Publisher, Switzerland (http://www.inderscience.com/ijfcm), Associate Editor of *Computational Methods in Structural Engineering, Frontiers in Built Environment*, and happens to be an Editorial Board member of *Springer Nature Applied Sciences*, *IGI Research Insights Books*, *Springer Book Series of Modeling and Optimization in Science and Technologies*, *Coupled Systems Mechanics (Techno Press)*, *Curved and Layered Structures (De Gruyter)*, *Journal of Composites Science (MDPI)*, *Engineering Research Express (IOP)*, and *Applications and Applied Mathematics: An International Journal.* He is also the reviewer of around 50 national and international journals of repute, and he was the President of the Section of Mathematical Sciences (including Statistics) of "Indian Science Congress" (2015–2016) and was the Vice President, "Orissa Mathematical Society" (2011–2013). Prof. Chakraverty is a recipient of prestigious awards, including the Indian National Science Academy (INSA) nomination under International Collaboration/Bilateral Exchange Program (with the Czech Republic), Platinum Jubilee ISCA Lecture Award (2014), CSIR Young Scientist Award (1997), BOYSCAST Fellow. (DST), UCOST Young Scientist Award (2007, 2008), Golden Jubilee Director's (CBRI) Award (2001), INSA International Bilateral Exchange Award ([(2010–2011) (selected but could not undertake), 2015 (selected)], Roorkee University Gold Medals (1987, 1988) for first positions in M.Sc. and M.Phil. (Comp. Appl.).

He is among the list of 2% of world scientists in Artificial Intelligence & Image Processing category based on a recent independent study (2020) done by Stanford University scientists. His world rank is 1862 out of 215114 researchers throughout the globe.

He has already guided 19 Ph.D. students and 12 are ongoing. Professor Chakraverty has undertaken around 16 research projects, totaling about Rs. 1.5 crores, as Principal Investigator funded by international and national agencies. He has brought in eight international students with different international/national fellowships to work in his group as PDFs, Ph.Ds, and visiting researchers for different periods. A good number of international and national conferences, workshops, and training programs have also been organized by him. His present research areas include Differential Equations (Ordinary, Partial, and Fractional), Numerical Analysis and Computational Methods, Structural Dynamics (FGM, Nano) and Fluid Dynamics, Mathematical and Uncertainty Modeling, Soft Computing and Machine Intelligence (Artificial Neural Network, Fuzzy, Interval and Affine Computations).

P. Karunakar, Assistant Professor at the Department of Mathematics in Anurag University, Hyderabad, has over 11 years of teaching and research experience. He got his Ph.D. in Mathematics from NIT Rourkela, Odisha in 2019. He received his M.Sc. in Applied Mathematics from NIT Warangal in 2009. Dr. Karunakar has published nine papers in reputed international journals. He is the co-author of a book entitled "Advanced Numerical Methods and Semi-Analytical Methods for Differential Equations" which was published by Wiley. He published four book chapters (two in a book published by Springer, one in a book by Wiley, and one in a book by Elsevier). Further, he has presented three papers in international conferences. He is serving as the Editorial Assistant of *International Journal of Fuzzy Computation and Modeling.*

https://doi.org/10.1142/9789811245367_fmatter

Contents

https://doi.org/10.1142/9789811245367_0001

Chapter 1

Numerical Aspects of Nonlinear Wave–Wave Interactions in Operational-Wave Models

G. Uma
Department of Ocean Engineering,
Indian Institute of Technology Madras, Chennai, India
umasathish82@gmail.com

Abstract

A wave model is used for the prediction of wind-generated waves in oceans. The wave information is necessary for the design and safety of many off-shore and near-shore structures and for the study of ocean surface–related physical processes. Currently, existing operational-wave models use third-generation wave model. There are several physical processes, described as source terms, available in these wave models. Of these source terms, the one due to nonlinear wave–wave interactions plays a vital role in operational third-generation wave models. The exact computation of nonlinear–linear energy transfer is time-consuming and challenging even with modern computers. The models either lack efficiency or accuracy. Therefore, its practical application still remains controversial for operational-wave models. This chapter provides a platform for better understanding of nonlinear wave–wave interactions and their sensitivity to water depth, directional distribution, and resolution.

Keywords: Spectral wave models, Haar wavelets, Numerical integration, Wave climate

1. Introduction: Discrete Spectral-Wave Models

The profile of ocean waves changes randomly with time. As a result, ocean waves are considered random or stochastic in nature. The evaluation of the properties of random waves is almost impossible on a wave-by-wave basis in time. It is thus natural to consider randomly changing waves as a stochastic process so that it is possible to evaluate the statistical properties of waves through the frequency and probability domains. In the stochastic approach illustrated by Ochi [1], ocean waves are categorized as a Gaussian random process for which the probability distribution of the wave profile obeys the normal probability law so that Gaussian closure hypothesis can be employed.

In describing random waves, the concept of wave (variance or energy) spectrum is introduced. The wave spectrum can be either the wave-number spectrum at a given time or the frequency spectrum at a given point. It gives full information about a time signal in the frequency domain and is represented by the Fourier transform of the auto-correlation of the time signal. Realistic spectra are usually based on measured data. It is thus a source of information from which the probabilistic prediction of various wave properties can be achieved in the probability domain. The prediction of wave information is necessary for design and safety of many off-shore and near-shore structures and for the study of ocean surface–related physical processes. In this relevance, a wave model is developed, which includes various physical processes that are responsible for the formation of wind-generated waves on the ocean surface.

Currently, third-generation wave models are used to predict the wave field. The third-generation Wave Model (WAM), widely used in many countries, adopts the Discrete-Interaction Approximation (DIA) method [2] to compute the quadruplet four-wave interactions. Implementation of DIA in third-generation wave models has greatly improved the response of wave models to complex variations of wind field compared with previous (first- and second-generation) wave models.

The first-generation wave models do not have an explicit expression for nonlinear four-wave interactions. The second-generation wave models use parametric methods for handling nonlinear four-wave interactions. The third-generation wave models use nonlinear four-wave interactions explicitly. All these wave models trace the development of each wave component in a gravity spectrum independently. Different wave components are coupled by a wave–wave interaction source term.

The deficiencies of first- and second-generation wave models and the crucial role of the nonlinear interactions in the SWAMP study [3] resulted in an explicit modeling of S_{nl} in wave models and thus the development of third-generation discrete spectral-wave models.

The aim of discrete spectral-wave models is to model each physical process with a separate source term, which may be difficult for many physical processes due to the following reasons:

(i) poor understanding of the underlying physics (such as in S_{wcap} term),
(ii) computational method is too time-consuming (as in S_{nl} term).

Formulation of these source terms are often based on a combination of theoretical studies and analysis of field and laboratory measurements.

The third-generation models compute the evolution of action density $N = N(t, x, y, k, \theta)$ as a function of time t, space variables x and y, wave number k, and direction θ. This evolution is described through the action-balance equation as follows:

$$\frac{\partial N}{\partial t} + \frac{\partial}{\partial x}(c_{g,x}N) + \frac{\partial}{\partial y}(c_{g,y}N) + \frac{\partial}{\partial k}(c_{g,k}N) + \frac{\partial}{\partial \theta}(c_{g,\theta}N) = S_{tot}, \tag{1}$$

where $N = E/\omega$, $E = E(t, x, y, k, \theta)$ is the energy-density spectrum and ω is the angular frequency. The first term in the left-hand side of this equation represents the local rate of change of action density in time, the second and third terms represent propagation of action in geographical space (with propagation velocity and in c_x and c_y in x- and y-spaces, respectively), and the fourth and fifth terms represent the propagation of action in spectral space (with propagation velocity c_k and c_θ and in k- and θ-spaces, respectively). The term on the right-hand side of this equation is the total source term, describing the changes in action density at each spectral component due to various physical processes. S_{tot} is given by

$$S_{tot} = \overbrace{\underbrace{S_{inp} + S_{wcap} + S_{nl}}_{deep\ water} + \{S_{fric} + S_{brk} + S_{tri} + S_{Bragg}\}}^{finite\ depth}, \tag{2}$$

where

- S_{inp} is the wind-generation input,
- S_{wcap} is the dissipation due to white capping, and
- S_{nl} is the nonlinear four-wave interactions exchanging wave action between a set of four waves.

The terms between the bracket are important in finite water:

- S_{fric} is the dissipation due to bottom friction,
- S_{brk} is the depth-limited wave breaking,
- S_{tri} is the nonlinear interactions between a set of three waves, and
- S_{Bragg} is the Bragg scattering term.

Of these source terms, the one due to nonlinear four-wave interactions, representing the energy exchange between the wave components, remain a key focus of research in the area of wave modeling. Recent literature reveals the vital role of this source term in the evolution (growth and decay) of wind-generated waves. The exact computation of S_{nl} involves the evolution of six-dimensional integral, which has an interaction function with singularities. Even with present-day computer technology and various improvements in the efficiency of computation of these integrals, it is still expensive for use in operational-wave models. Hasselmann [4] initially formulated the expression for S_{nl}. Later, he [5] himself came out with an exact solution, namely, EXACT-NL method for S_{nl}. Owing to its computational cost, its implementation in third-generation wave models became infeasible. Consequently, Hasselmann *et al.* [2] suggested a computationally efficient method, DIA. DIA is the most commonly used method in present third-generation wave models, WAVEWATCHIII [6], Simulating Waves Nearshore (SWAN) [7], WAM [8], and TOMAWAC [9]. DIA has several short comings which are reported in several works [10–13]. Apart from DIA, there exist certain exact methods [14–16]. Each exact method has its own deficiencies. Modifications to Webb's method were carried out by Tracy and Resio [17, 18] and are termed as Webb–Resio–Tracy (WRT) method. This chapter focuses on WRT method and proposes a modification to make it feasible for operational-wave models [13].

2. Mathematical Formulation: Nonlinear Wave–Wave Interactions Using Webb–Resio–Tracy Method

The nonlinear wave–wave interactions in the wave model proposed by Hasselmann[4],

$$\frac{\partial n_1}{\partial t} = \iiint G\left(\overrightarrow{k_1}, \overrightarrow{k_2}, \overrightarrow{k_3}, \overrightarrow{k_4}\right) \times \delta\left(\overrightarrow{k_1} + \overrightarrow{k_2} - \overrightarrow{k_3} - \overrightarrow{k_4}\right) \times \delta(W) \times D\left(\overrightarrow{k_1}, \overrightarrow{k_2}, \overrightarrow{k_3}, \overrightarrow{k_4}\right) d\overrightarrow{k_2}\, d\overrightarrow{k_3}\, d\overrightarrow{k_4}, \quad (3)$$

describes the rate of change of action density n_1 at a particular wave number $\overrightarrow{k_1}$ due to all the wave-resonating quadruplet interactions involved in it. $\delta(\cdot)$ and G denote the Dirac delta function and coupling coefficients, respectively. W is given by $W = \omega_1 + \omega_2 - \omega_3 - \omega_4$; here, ω_i is the angular frequency corresponding to the ith wave number k_i $(i = 1, \ldots, 4)$, and $n_i = n\left(\overrightarrow{k_i}\right)$ denote the action density related to the wave-number spectra by $n_i = \frac{F\left(\overrightarrow{k_i}\right)}{\omega_i}$. The expression for G is given by Heterich and Hasselmann [5] (finite depths) and Webb [15] (deep waters). The density term is given by $D\left(\overrightarrow{k_1}, \overrightarrow{k_2}, \overrightarrow{k_3}, \overrightarrow{k_4}\right) = [n_1 n_3(n_4 - n_2) + n_2 n_4(n_3 - n_1)]$.

The contribution for Eq. (3) is obtained from the vectors $\left\{\overrightarrow{k_2}, \overrightarrow{k_3}, \overrightarrow{k_4}\right\}$ interacting with $\overrightarrow{k_1}$. These vectors have to satisfy the following resonance conditions in terms of wave number and angular frequency, respectively:

$$\overrightarrow{k_1} + \overrightarrow{k_2} = \overrightarrow{k_3} + \overrightarrow{k_4} \quad (4)$$

and

$$\omega_1 + \omega_2 = \omega_3 + \omega_4. \quad (5)$$

The parameters ω and k are related linearly as

$$\omega^2 = gk\ \tanh(kd), \quad (6)$$

where g is the acceleration due to gravity and d is the water depth.

δ as a function of wave numbers in Eq. (3) is removed by integrating over $\overrightarrow{k_4}$ to obtain

$$\frac{\partial n_1}{\partial t} = \int T\left(\overrightarrow{k_1}, \overrightarrow{k_3}\right) d\overrightarrow{k_3}. \tag{6a}$$

The transfer integral $T\left(\overrightarrow{k_1}, \overrightarrow{k_3}\right)$ is given by

$$T\left(\overrightarrow{k_1}, \overrightarrow{k_3}\right) = \int G \times D\left(\overrightarrow{k_1}, \overrightarrow{k_2}, \overrightarrow{k_3}, \overrightarrow{k_4}\right) \times \delta\left(W\right) d\overrightarrow{k_2}. \tag{6b}$$

The first and foremost, also the most time-consuming part, in WRT method is to find the wave-resonating quadruplets $\left\{\overrightarrow{k_1}, \overrightarrow{k_2}, \overrightarrow{k_3}, \overrightarrow{k_4}\right\}$. This step is described in more detail in the following section.

Next, $\overrightarrow{k_2}$ is decomposed into normal and tangential components $(\overrightarrow{n}, \overrightarrow{s})$ as $\overrightarrow{k_{2,n}}$ and $\overrightarrow{k_{2,s}}$. With the help of δ property, the integration along $\overrightarrow{k_{2,n}}$ in Eq. (6b) is reduced to a closed integral as

$$T\left(\overrightarrow{k_1}, \overrightarrow{k_3}\right) = \oint G \times D\left(\overrightarrow{k_1}, \overrightarrow{k_2}, \overrightarrow{k_3}, \overrightarrow{k_4}\right) \times J \times ds. \tag{7}$$

Here, $J = |\overrightarrow{c_{g,2}} - \overrightarrow{c_{g,4}}|^{-1}$ is the Jacobian term with $\overrightarrow{c_{g,2}}$ and $\overrightarrow{c_{g,4}}$ denoting the group velocities of $\overrightarrow{k_2}$ and $\overrightarrow{k_4}$, respectively.

Equation (6a) is now expressed in polar coordinates as

$$\frac{\partial n\left(k_1, \theta\right)}{\partial t} = \iint T\left(\overrightarrow{k_1}, \overrightarrow{k_3}\right) k_3\, d\theta_3\, dk_3. \tag{8}$$

Following Tracy and Resio [17], the double integral in Eq. (8) is evaluated. However for finite depths, the scaling relations followed by the mentioned authors [17] cannot be used for wave numbers, coupling coefficients, and the Jacobian terms.

3. Multi-Resolution Analysis and Haar Wavelets

This section starts with a brief definition to multi-resolution analysis for Haar wavelets and proceeds with its application to the transfer integral, T.

3.1. *Multi-resolution Analysis or Multi-level Representation of Function*

The aim of multi-resolution analysis is to develop representation of a function $f(x)$ at various levels of resolution. To this end, we seek to expand $f(x) \in L^2(R)$ in terms of basis functions, called scaling function $\phi(x)$, and the wavelet function $\psi(x)$, which can be scaled to give the multiple resolution of the function.

A Multi-Resolution Analysis (MRA) of the set of square-integrable functions denoted by $L^2(R)$, equipped with the standard inner product $(\cdot,\cdot)$, is a chain of closed subspaces indexed by all integers

$$\ldots, V_{-1} \subset V_0 \subset V_1, \ldots$$

such that:

(i) $\bigcup_n V_n = L^2(R)$,
(ii) $\bigcap_n V_n = \{0\}$,
(iii) $f(\cdot) \in V_n \Leftrightarrow f(2\cdot) \in V_{n+1}$, and
(iv) $\{\phi(\cdot - k) : k \in Z\}$, constituting a complete orthonormal basis of V_0, where $\phi(\cdot)$ is a scaling function.

To obtain an MRA, it suffices to construct the scaling function $\phi(\cdot)$. The entire space chain can then be reconstructed from $\phi(\cdot)$ according to (iii) and (iv). Since $V_0 \subset V_1$ and from (iii) and (iv), it is easy to see that $\phi(\cdot)$ must be a linear combination of $\phi(\cdot - k) : k \in Z$, leading to the two-scale relation

$$\phi(\cdot) = 2\sum_{k \in Z} h_k \phi(2\cdot - k),$$

for a suitable set of coefficients $(\ldots, h_{-1}, h_0, h_1, \ldots)$.

Let W_0 denote the orthogonal complement of V_0 in V_1. A function $\psi(\cdot)$ whose integer translates to $\{\psi(\cdot - k) : k \in Z\}$ constituting an orthonormal basis of W_0 is called a wavelet. This wavelet function $\psi(\cdot)$ satisfies the two-scale relation

$$\psi(\cdot) = 2\sum_{k \in Z} g_k \psi(2x - k),$$

for a suitable set of coefficients $(\ldots, g_{-1}, g_0, g_1, \ldots)$. From (i)–(iv), it is clear that

$$\phi_{j,k}(\cdot) = 2^{j/2}\phi(2^j \cdot -k; j, k \in Z)$$

and

$$\psi_{j,k}(\cdot) = 2^{j/2}\psi(2^j \cdot -k; j, k \in Z)$$

form orthonormal bases of $L^2(R)$, where j and k are the scaling and translating parameters. For the well-known Haar wavelet, the scaling and wavelet functions are given by

$$\phi(x) = \begin{cases} 1, & 0 \leq x < 1, \\ 0, & \text{otherwise,} \end{cases}$$

$$\psi(x) = \begin{cases} 1, & 0 \leq x < \frac{1}{2}, \\ -1, & \frac{1}{2} \leq x < 1. \end{cases}$$

For any function $L^2(R)$, define P_J: $L^2(R) \longrightarrow V_J$ to be the projection of f onto the resolution space V_J by

$$f(x) = \sum_{k \in Z} c_{0,k}(f)\phi_{0,k}(x) + \sum_{0 \leq j \leq J} \sum_{\substack{0 \leq k \leq 2^j - 1 \\ k \in Z}} d_{j,k}(f)\psi_{j,k}(x), \tag{9}$$

where

$$c_{0,k} = \int f(x)\phi(x-k)\, dx \tag{10}$$

and

$$d_{j,k} = \int f(x)\psi_{j,k}(x)\, dx. \tag{11}$$

The parameter J denotes the maximum level of resolution of Haar wavelets to represent the function $f(x)$. The analysis of the function can be done using the scaling coefficients $c_{0,k}$ and the wavelet coefficients $d_{j,k}$. These coefficients are also called the average and detailed coefficients of the corresponding function, respectively.

3.2. *Numerical Integration Using Haar Wavelets and Their Application to Transfer Integral*

The evaluation of the definite integrals using Haar wavelets can be carried out by

$$\int_a^b f(x)\,dx = \frac{b-a}{2^{J+1}} \sum_{q=1}^{2^{J+1}} f(x_q), \tag{12}$$

with the Haar nodes given by $x_q = a + \frac{(b-a)(q-0.5)}{2^{J+1}}$; a and b are the lower and upper limits of the integration, respectively, and J represents the maximum level of resolution of Haar wavelets. More details can be found in [11].

An example for function approximation using Haar nodes is given in Figure 1. The function $sin^2 x$ is approximated at different levels of resolution, say $J = 2, 3$, and 4. Here, the values of J correspond to 8, 16, and 32 number of Haar nodes. Increasing the level of resolution approximates the exact graph more effectively.

Subsequent to our recent work [11], the Haar approximation of Eq. (7) yields

$$T\left(\overrightarrow{k_1}, \overrightarrow{k_3}\right) = \frac{L}{2^{J+1}} \sum_{k=1}^{2^{J+1}} G(s_k) \times D(s_k) \times J(s_k), \tag{13}$$

where s_k are the Haar nodes on the locus of $\overrightarrow{k_2}$, L is the length of the closed curve, and J is the maximum level of resolution.

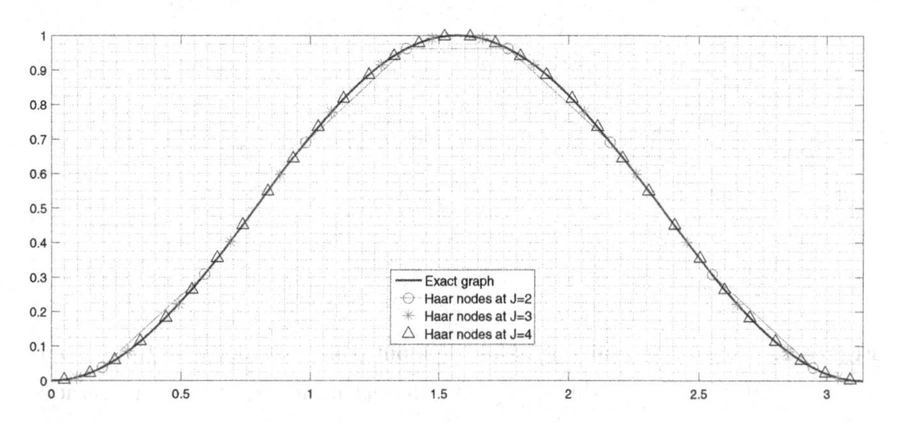

Figure 1. Function approximation of $sin^2 x$ using Haar wavelets.

3.3. *Procedure for Approximating the Locus Curve Using Haar Wavelets*

The curve Ω denoting $\overrightarrow{k_2}$ in the Cartesian plane is approximated by cubic-spline curves Γ: $(x(s), y(s))$, with s the cumulative chord length and $x(s)$ and $y(s)$ the cubic splines. The procedure is as follows.

Step 1: For the fixed $\overrightarrow{k_1}$ and $\overrightarrow{k_3}$ vectors, obtain the ordered set of points $\{x_i, y_i\}_{i=0}^{n-1}$ for $\overrightarrow{k_2}$ using explicit polar method presented by Vledder [19], yielding even number of points on the $\overrightarrow{k_2}$ and $\overrightarrow{k_4}$ vectors.

Step 2: Get $x(s)$ and $y(s)$ from Ω.
Assume s_i as the cumulative chord length at the ith point on Ω. For $\{s_i, x_i\}_{i=0}^{n-1}$ and $\{s_i, y_i\}_{i=0}^{n-1}$, get $x(s)$ and $y(s)$ separately. For the closed curve, s between from 0 to L.

Step 3: If s is chosen as the Haar node, the resulting points on the egg-shaped curve $\overrightarrow{k_2}$ will be equally spaced.

3.4. *Results of Haar-Wavelets Approximation to WRT (HWRT) Approximation on Locus Curves*

Figure 2 shows the approximation of locus curves using Haar wavelets at different depths. In all cases, the number of points on the locus curve is 32. Thus, the curve requires Haar approximation at the

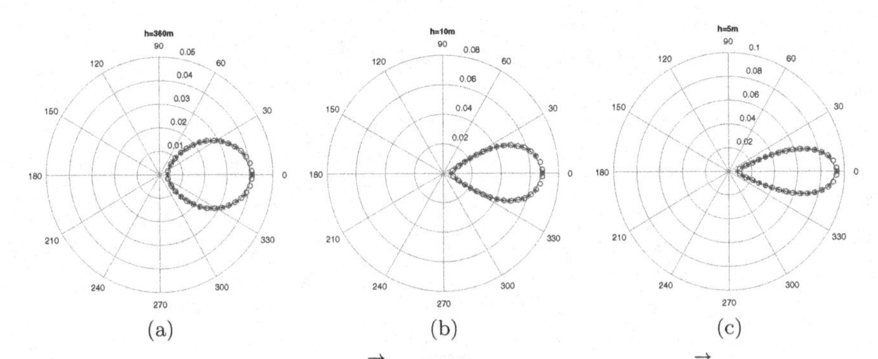

(a) (b) (c)

Figure 2. The locus results of $\overrightarrow{k_2}$ for the input pair of vectors $\overrightarrow{k_1} = (0.1339, 0^\circ)$ and $\overrightarrow{k_3} = (0.1449, 0^\circ)$ at different water depths using explicit polar method (line with stars) and present HWRT method (circles). $\overrightarrow{k_3}$ (square) always lies on $\overrightarrow{k_2}$.

fifth level of resolution. In Figure 2, orientations of input vectors are chosen to be the same. In the cases, the size of the locus curves grow bigger as the water gets shallower. These examples show the dependence of water depth on wave number. This impact of depth dependence on the wave numbers is reflected in the 1-D directionally integrated results of the nonlinear energy transfers.

4. Computational Results Using HWRT to Transfer Integral

In this section, a computational grid of 18 radian frequencies and 24 angles with 15° spacing ranging from $-180°$ to $180°$ is prepared for calculation. The tests in this section assume Pierson–Moskowitz (PM) spectra with anisotropic (two and eight powers of cosine distribution) cases of cosine distribution. The input spectra are shown in Figure 3. To show that the choice of spreading function also affects the shape and magnitude of the nonlinear transfers, we have considered the anisotropic cases for PM spectra. The input spectra considered in the present study correspond to published tests for the nonlinear transfer formulations. In order to test the sensitivity of nonlinear transfers to $\Delta\theta$, we present the nonlinear transfers as a function of $\Delta\theta$ for the input in Figure 3(a). The results are presented

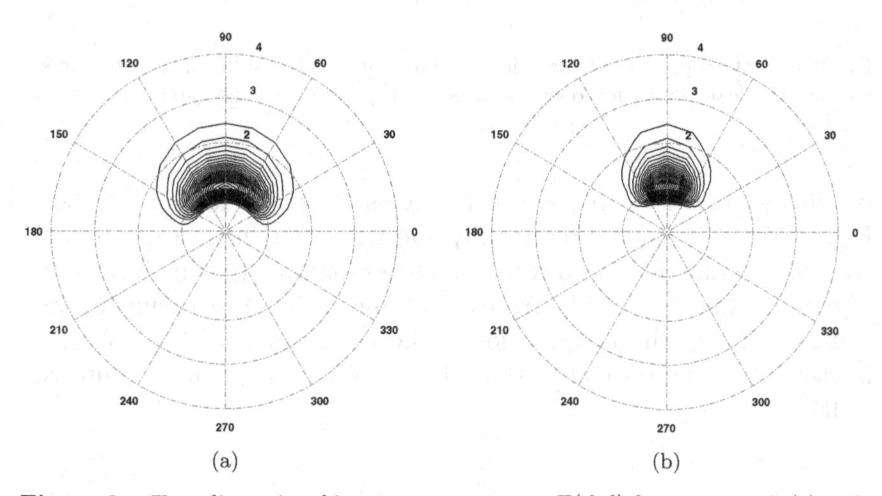

Figure 3. Two-dimensional input-energy spectra $E(f, \theta)$ for $\gamma = 1$, with (a) and (b) showing the anisotropic case of cosine spreading function.

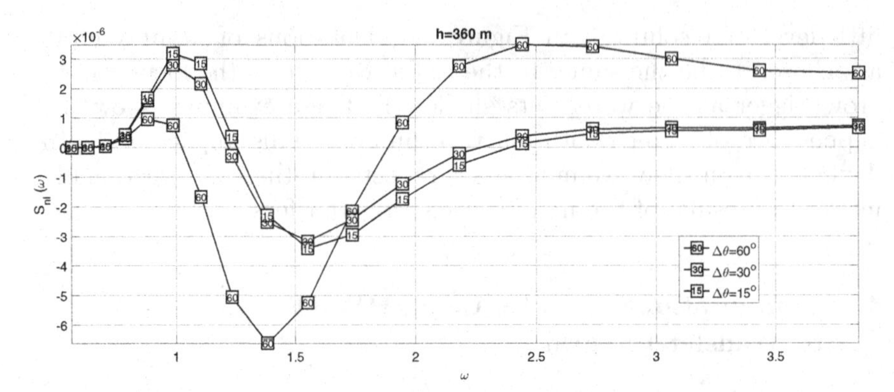

Figure 4. Input as in Figure 3(a) with S_{nl} and different angular resolutions $\Delta\theta = 60°$, $30°$, and $15°$.

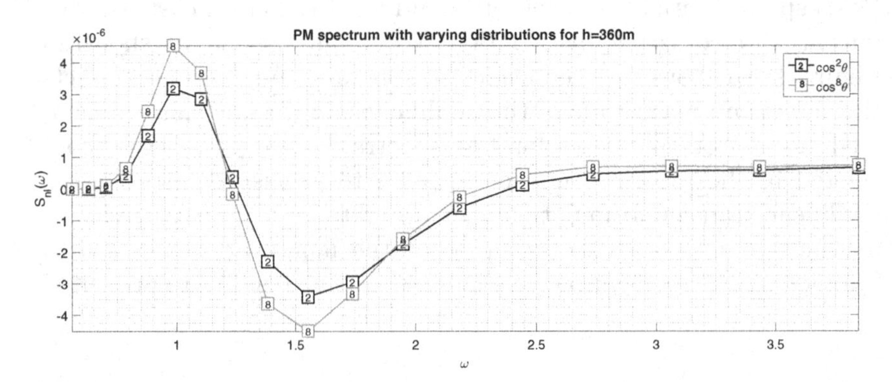

Figure 5. Comparison of directionally integrated 1-D nonlinear energy transfers for PM calculated for deeps waters corresponding to the input spectra in Figure 3.

in Figure 4, where minor sensitivity is observed for $\Delta\theta \leq 30°$. Figures 5 and 6 show directionally and frequency-integrated energy-transfer results for deep waters, corresponding to the inputs in Figure 3. The result illustrates that increasing the cosine power causes a rise in the magnitudes of the nonlinear transfers. Also, it is clear from the outcomes that the spreading function is centered at 90°.

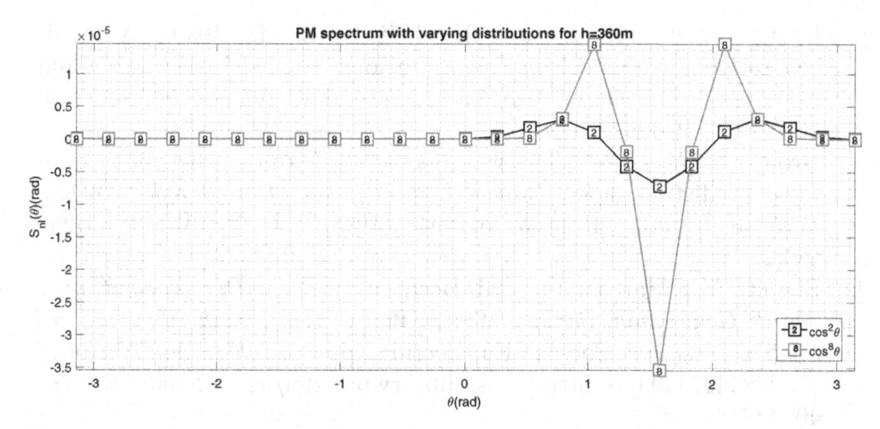

Figure 6. Comparison of frequency-integrated 1-D nonlinear energy transfers for PM calculated for deeps waters corresponding to the input spectra in Figure 3.

References

[1] Ochi, M. K. (1998). *Ocean Waves: The Stochastic Approach*, Cambridge Ocean Technology Series, Cambridge University Press. doi: 10.1017/CBO9780511529559.

[2] Hasselmann, K., Hasselmann, K., Allender, J. H., and Barnett, T. P. (1985). Computations and parameterizations of the nonlinear energy transfer in a gravity-wave spectrum. Part (ii): Parameterizations of the nonlinear energy transfer for application in wave models. *Journal of Physical Oceanography*, 15, 1378–1392. doi: 10.1175/1520-0485(1985)015⟨1378:CAPOTN⟩2.0.CO;2.

[3] SWAMP. (2021). *Ocean Wave Modeling*, 1st edn., Springer: USA.

[4] Hasselmann, K. (1960). Grundgleichungen der seegangsvoraussage. *Schiffstechnik*, 7, 191–195.

[5] Hasselmann, S. and Hasselmann, K. (1981). *A Symmetrical Method of Computing the Nonlinear Transfer in a Gravity Wave Spectrum.* Hamburger geophysikalische Einzelschriften/A: Wittenborn.

[6] Tolman, H. L. (1991). A third-generation model for wind waves on slowly varying, unsteady, and inhomogeneous depths and currents. *Journal of Physical Oceanography*, 21(6), 782–797. doi: 10.1175/1520-0485(1991)021⟨0782:ATGMFW⟩2.0.CO;2.

[7] Booij, N., Ris, R. C., and Holthuijsen, L. H. (1999). A third-generation wave model for coastal regions: 1. Model description and validation. *Journal of Geophysical Research*, 104, 7649–7666. doi: 10.1029/98JC02622.

[8] Group, T. W. (1988). The wam model — A third generation ocean wave prediction model. *Journal of Physical Oceanography*, 18(12), 1775–1810. doi: 10.1175/1520-0485(1988)018⟨1775:TWMTGO⟩2.0.CO;2.

[9] Benoit, M., Marcos, F., and Becq, F. (1996). *Development of a Third Generation Shallow-Water Wave Model with Unstructured Spatial Meshing*, Coastal Engineering, pp. 465–478. doi: 10.1061/9780784402429.037. https://ascelibrary.org/doi/abs/10.1061/9780784402429.037.

[10] Prabhakar, V. and Uma, G. (2016). A polar method using cubic spline approach for obtaining wave resonating quadruplets. *Ocean Engineering*, 111, 292–298. doi: 10.1016/j.oceaneng.2015.10.054. http://www.sciencedirect.com/science/article/pii/S0029801815006010.

[11] Uma, G., Prabhakar, V., and Hariharan, S. (2016). A wavelet approach for computing nonlinear wave–wave interactions in discrete spectral wave models. *Journal of Ocean Engineering and Marine Energy*, 2(2), 129–138. doi: 10.1007/s40722-015-0041-3.

[12] Uma, G., Vadari, P., and Sannasiraj, S. A. (2018). Hybrid functions for nonlinear energy transfers at finite depths. *Journal of Ocean Engineering and Marine Energy*, 4, 187–198. doi: 10.1007/s40722-018-0115-0.

[13] Uma, G. and Sannasiraj, S. A. (2021). Hybrid functions for nonlinear energy transfers in third-generation wave models: Application to observed wave spectra. In: Sundar, V., Sannasiraj, S. A., Sriram, V., and Nowbuth, M. D. (eds.), *Proceedings of the Fifth International Conference in Ocean Engineering (ICOE2019)*, Springer: Singapore, pp. 355–360.

[14] Masuda, A. (1980). Nonlinear energy transfer between wind waves. *Journal of Physical Oceanography*, 10(12), 2082–2093. doi: 10.1175/1520-0485(1980)010⟨2082:NETBWW⟩2.0.CO;2.

[15] Webb, D. J. (1978). Nonlinear transfer between sea waves. *Deep-Sea Research*, 25, 279–298.

[16] Lavrenov, I. V. (2001). Effect of wind wave parameter fluctuation on the nonlinear spectrum evolution. *Journal of Physical Oceanography*, 31, 861–873. doi: 10.1175/1520-0485(2001)031⟨0861:EOWWPF⟩2.0.CO;2.

[17] Tracy, B. and Resio, D. (1982). Theory and calculation of the nonlinear energy transfer between sea waves in deep water. WIS Report 11, U.S. Army Engineer Waterways Experiment Station.

[18] Resio, D. T. and Perrie, W. (1991). A numerical study of nonlinear energy fluxes due to wave–wave interactions. Part 1: Methodology and basic results. *Journal of Fluid Mechanics*, 223, 609–629. doi: 10.1017/S002211209100157X.

[19] van Vledder, G. P. (2006). The WRT method for the computation of non-linear four-wave interactions in discrete spectral wave models. *Coastal Engineering*, 53, 223–242. doi: 10.1016/j.coastaleng.2005.10.011.

https://doi.org/10.1142/9789811245367_0002

Chapter 2

Impact of Tropical Cyclones on the Ocean Surface Waves Over the Bay of Bengal

Mayuresh Vaze and Naresh Krishna Vissa*

Department of Earth and Atmospheric Sciences,
National Institute of Technology Rourkela,
Rourkela, Odisha 769008, India
**vissan@nitrkl.ac.in*

Abstract

Tropical cyclones are considered as one of the most intense case of air–sea interaction processes. Tropical cyclones receive energy from the upper ocean surface through exchanges of momentum and enthalpy fluxes. Ocean wind-generated surface gravity waves play a vital role in modulating the ocean surface conditions and momentum fluxes. Therefore, understanding of the ocean surface waves response during the passage of tropical cyclones would be beneficial for improved storm prediction. In the present study, an attempt has been made to quantify the characteristic of ocean surface waves during the passage of tropical cyclones. Here, we use the 27 years (1993–2019) of global wave reanalysis data sets and tropical cyclone track to develop composite maps of ocean surface waves with regard to the tropical cyclone intensity (maximum sustained wind speed) translation speed over the Bay of Bengal basin. Results suggest that transitional speed plays a vital role for the symmetric and antisymmetric pattern of the ocean surface waves. Findings from the present study are useful for the validation and improvement of state-of-the-art coupled-wave atmosphere models.

Keywords: Tropical cyclones, Bay of Bengal, Ocean Surface waves

1. Introduction

Oceans hold about 97% of the Earth's water content and occupy 70% of the Earth's surface. Oceans are the main component of the Earth's climate system, playing an important role in driving weather and climate systems. One of the most prevalent features of the ocean surface is the wind wave. Wind waves, generated by sustained winds flowing over the free surface of the ocean, vary in size from ripples (a few cm) to rogue waves (about 100 feet). Knowledge of the state of the ocean is crucial for the transportation, climate regulation, medicine, food chain, and economic purposes. Understanding the ocean waves is important for climate and weather prediction, shipping routing, and offshore industry. In the ocean basins, waves propagate by gaining energy obtained from the wind, and during its propagation, various processes play an important in the dissipation [1]. Near the surface, the wind blowing over the air–sea interface generates small-wavelength waves (wavelets) with time; these wavelets further grow by extracting energy from wind stress [2].

Waves and their breaking affect the energy exchanges between the ocean and atmosphere and in turn affects global climates. The condition of sea can be used to determine the intensity of storms and related climate patterns [3]. Storm surges are one of most dangerous phenomena causing damage to life and property in coastal regions. Coastal engineering requires assessment of loads that the waves induce upon constructions near coasts. Design of coastal structures, such as sea walls and sea dikes, built for protection of low-lying areas from floods can be improved with better understanding of wave climate. Meteorological tsunamis are a type of harbor oscillations, which are of the same temporal and spatial scales as a typical tsunami and can be equally damaging [4].

In the last decades, there has been increasing attention among engineers and scientists in obtaining long-term information on waves through ships, buoys, and satellite altimetry. Buoys are the most widely used systems for obtaining wave data. Buoys are floating devices which along with providing sea state data and facilitating safe navigation also serve various other purposes, including acting as mooring stations for ships. Radar, lasers, and radio altimetry are the techniques used in the recorders above surface and are mounted either on aircrafts or satellites. Among remote sensing instruments,

X-band marine radars are employed for collecting key information about the seas via sea surface images, which provide detailed information about the near-surface wind and wave [5]. High-frequency radar (HFR) systems installed along coasts can be used to measure ocean wave directional spectrum up to 100 km offshore [6].

Numerical modelling of ocean waves hindcast and forecast has several applications, such as construction and management of offshore activates and developments, naval and harbor structures. Initially, Sverdrup and Munk developed wave prediction based on a statistical model using significant wave height [7]. Numerical wave models have been developed over the years to predict the behavior of waves and incorporate them in coupled ocean-wave–atmosphere models. There was rapid advancement in spectral wave modeling after Pierson [8] introduced the concept of wave spectrum, followed by the Joint North Sea Wave Project experiment in 1973. Since then, three generations of wave models have been developed, with every generation of models treating the nonlinear interaction term in a different manner. The study of surface waves has been done using third-generation models (WAM, WAVEWATCH III, SWAN, and UMWM). Several studies have shown that these models show discrepancies in the estimation of significant wave height, which has been attributed to the overestimation of the drag coefficients [2, 9]. Improvements in the parametrization of these exchange coefficients will ultimately lead to better prediction of Tropical Cyclones (TCs). Spectral wave models have existed for decades now, and the only improvement in results has been due to the improvement in the resolution and quality of data and advancements in parametrization of physical processes.

Tropical cyclones, one of the most dangerous phenomena which cause massive damage to life and property every year, have their origin over warm tropical waters. Hence, the accuracy of TC predictions plays a major role in minimizing the damage. TC associated storm surge and wind waves are the modulation of near-surface winds over the ocean surface and can be very hazardous to coastal regions because of the abnormal rise in water level [2, 10]. Improvements in TC prediction depend on improvements in the determination of the factors that affect TC evolution, track, and intensity. One of the most critical factors on which the intensity of TCs depends is the air–sea energy exchanges [11, 12]. The interaction between ocean and atmosphere occurs through surface gravity waves. Thus, the structure of the TC-generated waves and their interaction with the atmosphere in turn determines the development of the TC. Studies have been conducted over the years trying to understand the factors affecting the TC intensity. Momentum transfer is crucial for the air–sea exchange process; drag coefficient (C_D) is one of the important parameters that primarily depends on wind shear stress. Previous studies have shown that C_D increases monotonically with increasing near-surface wind; however, at extreme wind conditions, C_D reaches saturation (10 m height) [13, 14]. Holthuijsen *et al.* [15] studied the impact of breaking waves and the formation of whitecaps, trails, and white-outs on C_D. Earlier studies suggest that rapidly intensifying TCs are associated with younger surface waves [16]. Further, it was also found that "Goldilocks tropical cyclones," which moderate translation speed, have a higher tendency for intensification than other types of TCs. Saturation of large waves was found in the right-front quadrant of TCs with moderate to high wind speeds (called trapped-fetch waves) as a result of approximate resonance between the translation speed of TCs and wave group speed [17]. Studies have suggested that this saturation might be leading to the leveling off and decrease in C_D.

The east coast of India experiences landfall from tropical cyclones every year, and the intensity has been increasing in the past decade [18]. Warmer sea surface temperatures observed in the Bay of Bengal in the recent decade are conducive to cyclogenesis formation. Study of cyclones in the Bay of Bengal is thus important to save lives and to plan better mitigation strategies. Studies on different cyclones over the Bay of Bengal have shown that the use of coupled wave?hydrodynamic models for simulation of storm surge and wave characteristics gave better estimates of significant wave height and surge residuals [2, 18–20]. A Recent study on the Bay of Bengal post-monsoon Hudhud cyclone (2014) suggests that the wave height is higher on the left side of the storm, whereas the observed currents are propagating in the opposite direction [21]. In the present study, an attempt has been made to quantify and understand the TC-induced wind waves variability and its relationship with the storm intensity and translation speed over the Bay of Bengal.

2. Data and Methods

The position, maximum sustained wind speed, radius of maximum winds, eye diameter, and other related information for a TC are obtained from the Joint Typhoon Warning Center (JTWC) best track archive of the Bay of Bengal for the period (1993–2019) (https://www.metoc.navy.mil/jtwc/jtwc.html). Over the Bay of Bengal, the TCs show a bimodal character during pre-monsoon (March–May) and post-monsoon (October–December) seasons. Whereas, during the Indian summer monsoon, in the presence of a large vertical shear, the TC activity is inhibited [22]. In the present study, TC tracks are segregated to pre-monsoon and post-monsoon seasons. The TCs are further categorized under levels 1–5, according to the Saffir–Simpson hurricane wind scale (https://www.nhc.noaa.gov/aboutsshws.php). Details of the Saffir–Simpson hurricane winds scale are given in Table 1. Based on the translation speed (U_h) of the storm, TCs are grouped into slow ($0 \leq U_h \leq 3\,\mathrm{m\,s^{-1}}$), medium ($3 < U_h \leq 7\,\mathrm{m\,s^{-1}}$), and fast ($U_h > 7\,\mathrm{m\,s^{-1}}$) [16].

The near-surface ocean winds of Cross-Calibrated Multi-Platform (CCMP) wind vector analysis product are used and were obtained from remote sensing systems (http://www.remss.com/measurements/ccmp/). This CCMP product comprises of wind data from

Table 1. Tropical cyclone categories and the corresponding range of maximum wind speed as per the Saffir–Simpson hurricane wind scale.

Category	Maximum Sustained Winds (kt)	Damage
1	64–82	Damage may occur due to very dangerous winds
2	83–95	Extensive damage could happen in association with extremely high wind speeds
3	96–112	Devastating damage could happen in association with extremely high wind speeds
4	113–136	Catastrophic damage would occur
5	>136	Catastrophic damage would occur

RSS radiometer, QuikSCAT, *in situ* (moored buoy wind data), ASCAT scatterometer wind vectors, and ERA-interim model wind. Variational analysis method is used to produce the daily maps with spatial resolution of 0.25° × 0.25°. For the study period, the significant wave height data of Global Ocean Waves Reanalysis WAVERYS are obtained from the Copernicus Marine Services Information (https://marine.copernicus.eu/). The WAVERYS data is available at daily interval with spatial resolution of quarter degree. Using the near-surface wind and significant wave height, the role of TC-induced wave variations are assessed by diagnosing various variables, such as inverse wave age (Ω_n) and phase speed (C_p).

The empirical formula for calculating inverse wave age is given by [16]

$$\Omega_n = 0.18 \times \eta_n^{-0.29},$$

where Ω_n is the inverse wave age and η_n is the dimensionless wave height. Here, the dimensionless wave height is given by

$$\eta_n = \left(\frac{C_{H_s}}{2U_{10}}\right)^4, \quad \text{where } C_{H_s} = \sqrt{gH_s}.$$

Also, the inverse wave age can be calculated from phase speed (C_p) using the empirical formula

$$C_p = 2C_g = 4.6459 * \frac{H_s^{0.5848}}{U_{10}^{0.1696}},$$

$$\Omega_n = U_{10}C_p^{-1},$$

where g is the acceleration due to gravity, H_s is the significant wave height, and U_{10} is the 10 m wind speed.

3. Results and Discussion

3.1. *Climatology*

The significant wave height climatology for the pre-monsoon and post-monsoon seasons along with the TC tracks are shown in Figure 1. The mean significant wave height of the north Indian ocean

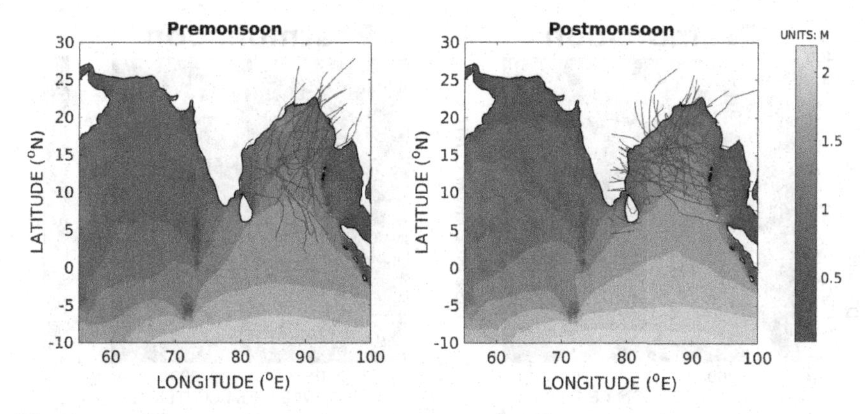

Figure 1. Pre-monsoon and post-monsoon significant wave height climatology (1993–2019) and the corresponding TC tracks.

for pre-monsoon and post-monsoon vary from 0.5 to 2 m; the seasonal mean is higher for post-monsoon compared to that for the pre-monsoon season. In the north Indian ocean, higher (lower) values are observed in the eastern (western) equatorial ocean and the Bay of Bengal (Arabian sea). Higher significant wave heights during post-monsoon in the Bay of Bengal are associated with the Southern ocean swells [23]. Over the southern Indian ocean, the maximum sustained winds increased due to global warming, and it has a large impact on the Southern ocean swells and influencing the waves elsewhere [24]. The occurrence of TC are highest post monsoon, during which they are mostly propagating to move in the westward and north-westward directions, whereas during the pre-monsoon season, most of the TC are moving in the north and north-eastward directions.

3.2. *TC-Induced Significant Wave Height Variations*

The composites of significant wave height and near-surface wind for the pre-monsoon and post-monsoon cyclones are shown in Figure 2. The composites of TC-centered domain are rotated with direction of storm heading by the arrow. For both pre-monsoon and post-monsoon, the TC-induced winds and significant wave height show asymmetric pattern. Higher winds and significant wave height are observed on the right side of the storm. Maximum winds are observed on the right-backward side of the storm, whereas the maximum significant wave height is noticed on the right-forward direction.

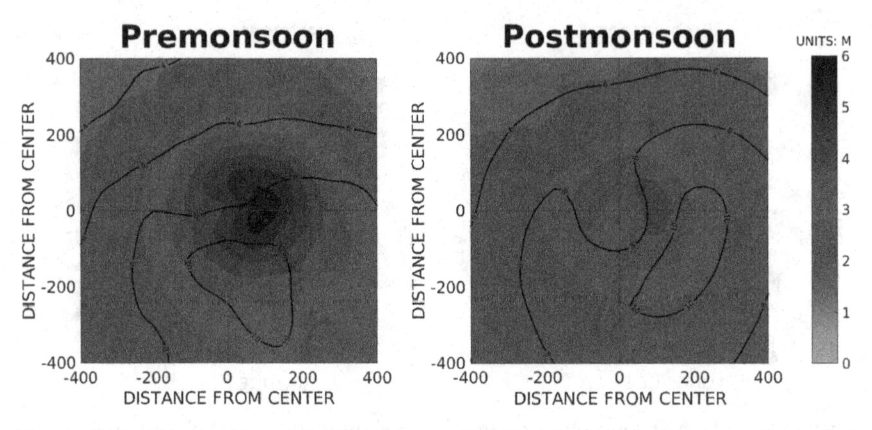

Figure 2. Composites of significant wave height (shaded) and near-surface wind (contour) within ~4° radius of the TC center for pre-monsoon and post-monsoon.

The relationship between TC intensity and significant wave height is shown in Figure 3. At lower TC intensity (C1), the asymmetry of significant wave height is low, whereas with increasing intensity, the asymmetry of the winds is increasing. Consistent with the wave heights, winds are also higher in the right side (rearward) of the storm. The influence of the TC winds on waves are significantly high during post-monsoon season. The results are consistent with the findings of [16, 17, 25]. The relationship between the significant wave height induced by the TC translation speed and winds is shown in Figure 4. For slow moving TCs, winds and significant wave height show a symmetric nature, more significantly for post-monsoon season. With the increasing speed of the TC, asymmetry of the waves and winds are prominent. However, fast-moving TCs are evidently not observed during the pre-monsoon season.

The composites of inverse wave age of different TC intensities for pre-monsoon are shown in Figure 5. For low-intensity TCs (categories 1 and 2), higher values of inverse wave age are evident in the rear side of the TC center. Higher values of wave age (>1) signify young waves. However, with increasing TC intensity, young waves are observed in the right-forward quadrant. The composites of TC-induced wave age for post-monsoon under different TC categories are shown in Figure 6. In contrast to the pre-monsoon season, significant young waves are observed in the right-forward side of the

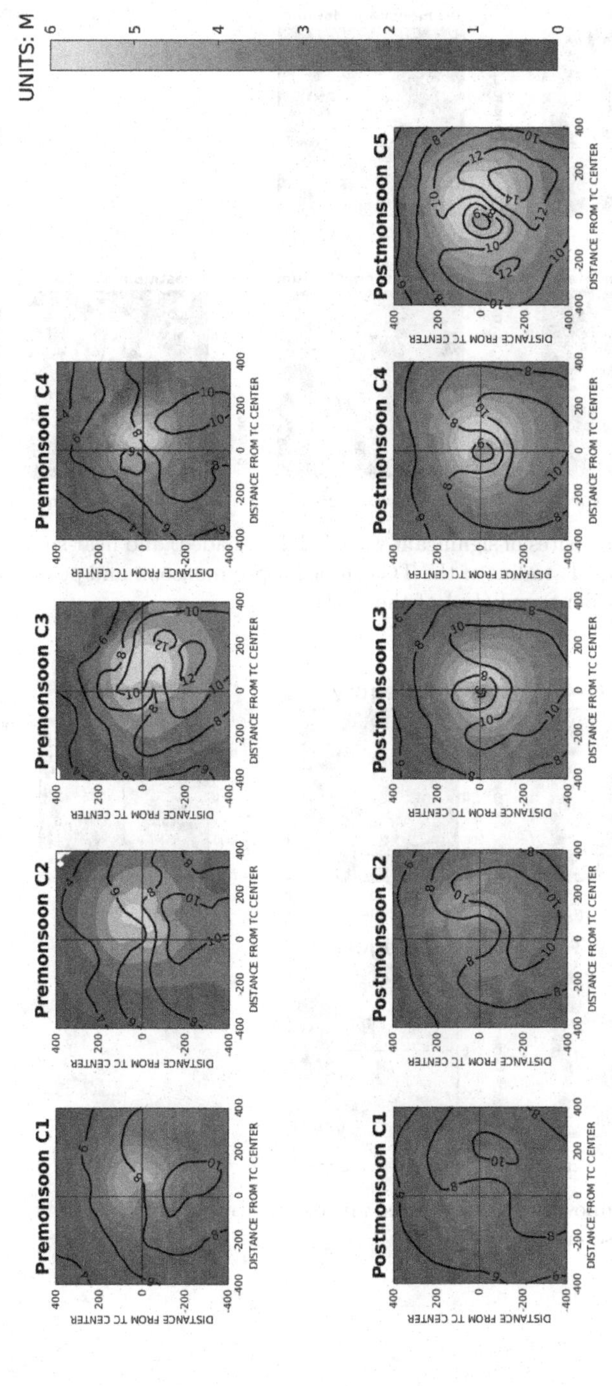

Figure 3. Composites of significant wave height (shaded) and near-surface wind (contour) within -4° radius of the TC center for pre-monsoon and post-monsoon based on intensity of the storm.

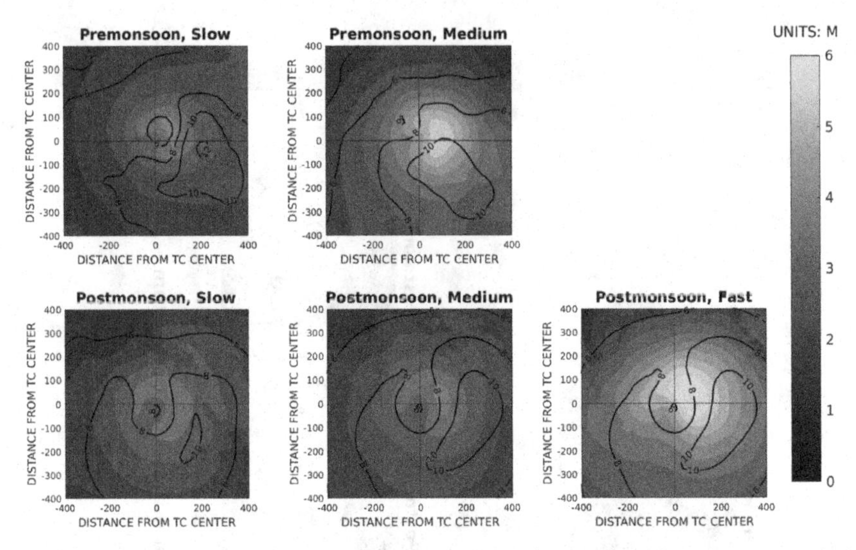

Figure 4. Composites of significant wave height (shaded) and near-surface wind (contour) within $-4°$ radius of the TC center for pre-monsoon and post-monsoon based on translation speed (U_h) of the storm.

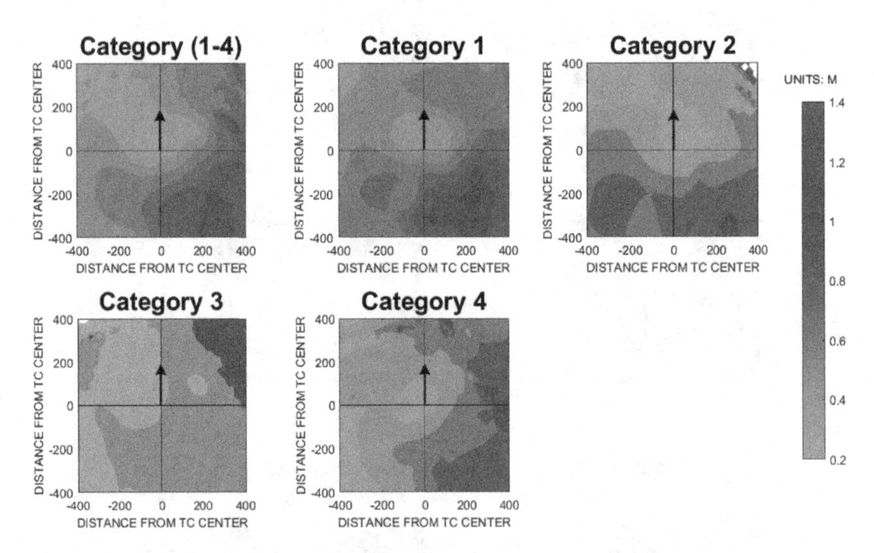

Figure 5. Composites of TC-induced inverse wave age of different TC categories for the pre-monsoon season.

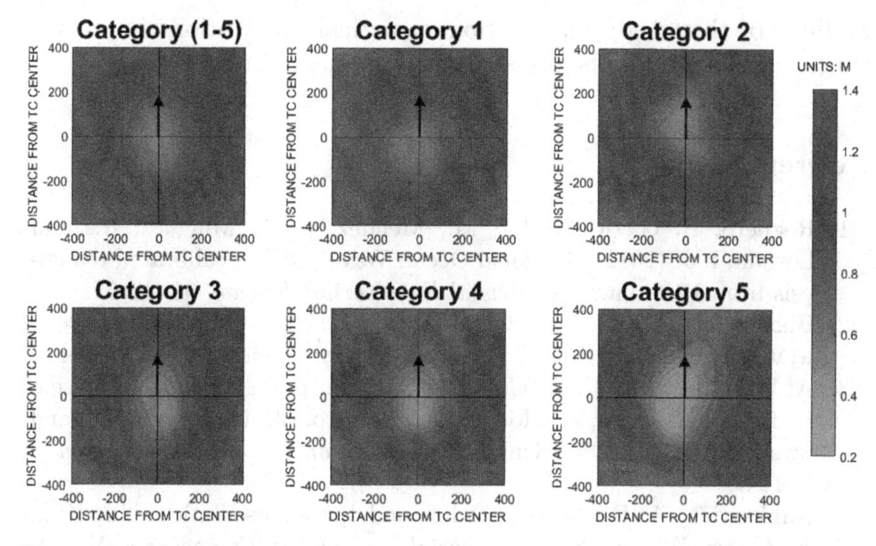

Figure 6. Composites of TC-induced inverse wave age of different TC categories for the post-monsoon season.

TC center. With higher-intensity TCs, symmetry of young wave is observed along the TC track.

4. Conclusions

The evolution of TC-induced significant wave height characteristics and the influence of storm intensity and translation speed are examined for the pre-monsoon and post-monsoon storms over the Bay of Bengal. With the increase of TC intensity, high energy waves are observed along the right-backward side of the storm; however, with low-intensity, symmetric wave properties are observed for the pre-monsoon and post-monsoon seasons. The influence of TC-induced wave characteristics are dominant for the pre-monsoon season. Slow response of post-monsoon TC to the wave energy might be associated with strong saline stratification. Further, TC-induced inverse wave age properties are assessed in the present study; the results reveal that young waves are more significantly evident for higher-intensity TCs. The young waves are evident along both sides of TC track, unlike the wind pattern. This could be associated to the storm speed being much higher than the group speed of the wind waves [25]. The findings of this study have various practical applications in storm-surge modeling and offshore structures management.

References

[1] Reguero, B. G., Menéndez, M., Méndez, F. J., Mínguez, R., and Losada, I. J. (2012). A Global Ocean Wave (GOW) calibrated reanalysis from 1948 onwards. *Coastal Engineering Journal*, 65, 38–55.
[2] Bhaskaran, P. K. (2019). Challenges and future directions in ocean wave modeling a review. *Journal of Extreme Events*, 6(2), 1950004.
[3] Ardhuin, F., Stopa, J. E., Chapron, B., Collard, F., Husson, R., Jensen, R. E., Johannessen, J., Mouche, A., Passaro, M., Quartly, G. D., and Swail, V. (2019). Observing sea states. *Frontiers in Marine Science*, 6, 124.
[4] Rabinovich, A. B. (2010). Seiches and harbor oscillations. In: Kim, Y. C. (eds.), *Handbook of Coastal and Ocean Engineering*, Vol. 1, World Scientific, pp. 193–236.
[5] Huang, W., Liu, X., and Gill, E. W. (2017). Ocean wind and wave measurements using X-band marine radar: A comprehensive review. *Remote Sensing*, 9, 1261.
[6] Wyatt, L. R., Green, J. J., and Middleditch, A. (2011). HF radar data quality requirements for wave measurement. *Coastal Engineering Journal*, 58(4), 327–336.
[7] Thomas, T. J. and Dwarakish, G. S. (2015). Numerical wave modelling — A review. *Aquatic Procedia*, 4, 443–448.
[8] Pierson Jr., W. J. (1955). Wind generated gravity waves. In: Landsberg, H. E. (eds.), *Advances in Geophysics*, Vol. 2, Elsevier, New York, NY, pp. 93–178.
[9] Fan, Y., Hwang, P., and Yu, Y. (2020). Surface gravity wave modeling in tropical cyclones. In: Essa, K.S., Di Risio, M., Celli, D., and Pasquali, D. (eds.), *Geophysics and Ocean Waves Studies*, IntechOpen, London, UK, p. 1.
[10] Bhaskaran, P. K., Gayathri, R., Murty, P. L. N., Bonthu, S., and Sen, D. (2014). A numerical study of coastal inundation and its validation for Thane cyclone in the Bay of Bengal. *Coastal Engineering*, 83, 108–118.
[11] Vissa, N. K., Satyanarayana, A. N. V., and Prasad Kumar, B. (2012). Response of Upper Ocean during passage of MALA cyclone utilizing ARGO data. *International Journal of Applied Earth Observation and Geoinformation*, 14, 149–159.

[12] Vissa, N. K., Satyanarayana, A. N. V., and Kumar, B. P. (2013). Intensity of tropical cyclones during pre-and post-monsoon seasons in relation to accumulated tropical cyclone heat potential over Bay of Bengal. *Natural Hazards*, 68(2), 351–371.

[13] Powell, M. D., Vickery, P. J., and Reinhold, T. A. (2003). Reduced drag coefficient for high wind speeds in tropical cyclones. *Nature*, 422(6929), 279–283.

[14] Takagaki, N., Komori, S., Suzuki, N., Iwano, K., and Kurose, R. (2016). Mechanism of drag coefficient saturation at strong wind speeds. *Geophysical Research Letters*, 43(18), 9829–9835.

[15] Holthuijsen, L. H., Powell, M. D., and Pietrzak, J. D. (2012). Wind and waves in extreme hurricanes. *Journal of Geophysical Research: Oceans*, 117(C9).

[16] Zhang, L. and Oey, L. (2019). An observational analysis of ocean surface waves in tropical cyclones in the Western North Pacific Ocean. *Journal of Geophysical Research: Oceans*, 124(1), 184–195.

[17] Bowyer, P. J. and MacAfee, A. W. (2005). The theory of trapped-fetch waves with tropical cyclones-An operational perspective. *Weather and Forecasting*, 20, 229–244.

[18] Murty, P. L. N., Bhaskaran, P. K., Gayathri, R., Sahoo, B., Kumar, T. S., and SubbaReddy, B. (2016). Numerical study of coastal hydrodynamics using a coupled model for Hudhud cyclone in the Bay of Bengal. *Estuarine, Coastal and Shelf Science*, 183, 13–27.

[19] Bhaskaran, P. K., Nayak, S., Bonthu, S. R., Murty, P. N., and Sen, D. (2013). Performance and validation of a coupled parallel ADCIRC–SWAN model for THANE cyclone in the Bay of Bengal. *Environmental Fluid Mechanics*, 13, 601–623.

[20] Murty, P. L. N., Sandhya, K. G., Bhaskaran, P. K., Jose, F., Gayathri, R., Nair, T. B., Kumar, T. S., and Shenoi, S. S. C. (2014). A coupled hydrodynamic modeling system for PHAILIN cyclone in the Bay of Bengal. *Coastal Engineering Journal*, 93, 71–81.

[21] Samiksha, V., Vethamony, P., Antony, C., Bhaskaran, P., and Nair, B. (2017). Wave–current interaction during Hudhud cyclone in the Bay of Bengal. *Natural Hazards and Earth System Sciences*, 17(12), 2059–2074.

[22] Akter, N. (2015). Mesoscale convection and bimodal cyclogenesis over the Bay of Bengal. *Monthly Weather Review*, 143, 3495–3517.

[23] Anoop, T. R., Kumar, V. S., Shanas, P. R., and Johnson, G. (2015). Surface wave climatology and its variability in the North Indian Ocean based on ERA-Interim reanalysis. *Journal of Atmospheric and Oceanic Technology*, 32(7), 1372–1385.

[24] Gupta, N., Bhaskaran, P. K., and Dash, M. K. (2015). Recent trends in wind-wave climate for the Indian Ocean. *Current Science*, 108(12), 2191–2201.

[25] Nair, M. A., Kumar, V. S., and George, V. (2021). Evolution of wave spectra during sea breeze and tropical cyclone. *Ocean Engineering*, 219, 108341.

https://doi.org/10.1142/9789811245367_0003

Chapter 3

Theoretical and Numerical Studies of Boussinesq Equations for Onshore Shallow-Water Wave Propagation

Prashant Kumar*, Vinita and Rajni

Department of Applied Sciences,
National Institute of Technology Delhi,
Delhi 110040, India
**prashantkumar@nitdelhi.ac.in*

Abstract

Boussinesq Equations (BEs) are defined to model and determine the nonlinear transformation of shallow-water surface waves induced by shoaling, diffraction, and partial reflection of interior boundaries. In this chapter, one-dimensional (1-D) and two-dimensional (2-D) analytical and numerical solutions of BEs are determined utilizing the concept of shallow-water waves over a variable water depth. The analytical solutions of the 1-D and 2-D Boussinesq-type equations are defined by first-integral method. Linear dispersion properties of BEs are also discussed. The numerical solution of 1-D BEs for regular and irregular waves is estimated by using the finite-difference method with Crank–Nicolson procedure. The numerical model is validated by using the analytical and experimental studies for shallow-water waves over the constant slope. The wave spectrum is also determined by using Pierson–Moskowitz spectra for shallow-water waves. The proposed numerical model is also applicable to any realistic domain for practical applications involving coastal regions.

Keywords: Boussinesq equations, Shallow water waves, Finite difference method, Wave spectrum

1. Introduction

Over the past decades, Boussinesq Equations (BEs) have been extended and shown to be efficient in modeling free-surface elevation and water-wave generation in deep to shallow waters over varying bathymetry with existing boundary conditions. BEs are applied to provide precise description of the surface of water waves in coastal areas. The spread of superficial gravitational waves over changing topography is an important problem for coastal engineers. The propagation of waves near the coast consists of nonlinear physical phenomena of water waves with wave characteristics. Boussinesq wave theory emulates highly diffuse nonlinear waves, their variable depth velocities, and simulates long-wave currents which are deeper, but still less than the depth of the surf zone near the shoreline. The efficiency of precisely forecast wave changes from deep to shallow water is important for understanding wavefront processes. Whereas, shallow water waves are considered propagating waves without changes in phase speed and ripples. In addition, the free surface height increases because of the seabed in the shallow water zone.

Initially, the BEs represent an inviscid depth-averaged approximation of the Navier–Stokes equations. Peregrine defined classical BE to include only weak dispersion and weak nonlinearity, and these equations are valid only for simulating long waves in shallow waters [1]. The enhanced dispersion and shoaling gradients of the new set of BEs are defined for separate bathymetry gradually with depth-integrated velocity variables [2–4]. The improved Boussinesq model equations have been defined by Nwogu, who developed the arbitrary depth velocity for variable water depth [5]. BEs are considered with considerable accuracy by including dispersion properties [3, 6, 7]. However, the Boussinesq model incorporates wave breaking and shorelines to represent new surf-zone models that represent waves and wave-simulate currents, containing wave setup, speed currents, and longshore flows [8–13]. Each model includes different form arrangements of dispersion terms. There are other higher-order Boussinesq models described with preferable distribution relation and more suitable nonlinearity [6, 14–16].

The numerical solution of the 1-D BEs is defined by higher-order finite-difference numerical methods with predictor–corrector scheme for time and space discretizations [7]. The 2-D BEs of nonlinear wave propagation and transformation in near-shore zones have been defined by Karambas [17]. Numerical methods have been developed to investigate water wave disturbance and hydrodynamics of swash and surf zones [18–24]. The improvement of the 2-D wave model based on the extended BEs has been defined by Chuanjian [19]. Analytical methods have been defined to solve Boussinesq equations by some researchers [25–27]. Several mathematical models have been built to incorporate the effects of refraction, diffraction, and constant and variable bathymetry [8, 11, 12, 23, 28–34]. Accurate analysis of the wave effects is accomplished with nonlinear wave transformations, such as wave breaking and shoaling, and harmonic interaction has been described better in fully nonlinear BEs [35–37]. The numerical solution of 2-D extended BEs has been defined by using a hybrid numerical scheme [38]. The solution of improved BEs with a meshless finite-difference method has been defined by Zhang [39]. Moreover, the trough instabilities in Boussinesq formulations for water waves is defined by Madsen [40], and the numerical observation of harbor fluctuations is induced by focus transient wave groups [41].

In this study, the Finite-Difference Method (FDM) has been used to solve BEs with staggered-grid system and Crank–Nicolson scheme of implicit form used with forward time and central space of the FDM. The analytical solutions of 1-D and 2-D BEs are defined by using the first-integral method. The dispersion characteristics of the BEs are described, which is suitable for a wider range of water depths. The PM spectrum analysis is defined for water-wave propagations. The numerical results are validated with existing literature and experimental data. The results of the present numerical model is applicable to coastal regions with sloping bathymetry.

2. Mathematical Formulation

Assume the fluid to be non-compressible and inviscid and fluid flow to be non-rotational. The mathematical model is developed by using the incompressible Euler's equations of motion and continuity equations. Cartesian coordinates (x, z) are considered, where x axis is representing the horizontal direction and z axis is along the vertically upwards direction. Here, $\xi(x, y, t)$ refer to water-surface elevation, L is wavelength, the wave height is H, h_0 denotes depth of the water wave from still-water level to bottom, and $u(x, y, t)$ represents the velocity variable of the water waves in x direction over the depth of water $h(x, y)$.

2.1. *Derivation of Nonlinear Boussinesq Equations (NBEs)*

The dimensionless system of variables denoted by symbols x, y, z, and t are written as follows:

$$x = \frac{X}{L}, \quad y = \frac{Y}{L}, \quad z = \frac{Z}{h_0}, \quad t = \frac{\sqrt{gh_0}}{L}T. \tag{1}$$

The expressions of the dimensionless velocity vectors u, v, and w are given in the form of dimensional velocity vectors as follows:

$$\begin{aligned} u &= \frac{h_0}{A\sqrt{gh_0}}U, \quad v = \frac{h_0}{A\sqrt{gh_0}}V, \quad w = \frac{h_0^2}{AL\sqrt{gh_0}}W, \\ h &= \frac{H}{h_0}, \quad \xi = \frac{\xi^*}{A}, \quad p = \frac{P}{\rho g A}, \end{aligned} \tag{2}$$

where P is the dimensional form of pressure. Here, ξ^* refers to surface elevation in the dimensional form, and A represents wave amplitude. The continuity equation is written in dimensionless form as

$$\varepsilon^2\left(\frac{\partial u}{\partial x} + \frac{\partial v}{\partial y}\right) + \frac{\partial w}{\partial z} = 0. \tag{3}$$

The Euler's equations of motion are written in dimensionless form as

$$\varepsilon^2\frac{\partial u}{\partial t} + \varepsilon^2\psi\left(u\frac{\partial u}{\partial x} + v\frac{\partial u}{\partial y}\right) + \psi w\frac{\partial u}{\partial z} + \varepsilon^2\frac{\partial p}{\partial x} = 0, \tag{4}$$

$$\varepsilon^2\frac{\partial v}{\partial t} + \varepsilon^2\psi\left(u\frac{\partial v}{\partial x} + v\frac{\partial v}{\partial y}\right) + \psi w\frac{\partial v}{\partial z} + \varepsilon^2\frac{\partial p}{\partial y} = 0, \tag{5}$$

$$\psi\frac{\partial w}{\partial t} + \psi^2\left(u\frac{\partial w}{\partial x} + v\frac{\partial w}{\partial y}\right) + \frac{\psi^2}{\varepsilon^2}w\frac{\partial w}{\partial z} + \psi\frac{\partial p}{\partial z} + 1 = 0. \tag{6}$$

The parameter $\psi = A/h_0$ is a measure of nonlinearity and $\varepsilon = h_0/L$ measures frequency dispersion. The non-rotational condition is given by $\nabla \times (u, v, w) = 0$. The boundary conditions for the fluid motion are as follows:

(1) Free-surface dynamic boundary condition is given as

$$p = p_{atm} \quad at\ z = \psi\xi(x, y, t), \tag{7}$$

where p_{atm} is the atmospheric pressure.

(2) Free-surface kinematic boundary condition is expressed as

$$w = \varepsilon^2 \left(\frac{\partial \xi}{\partial t} + \psi u \frac{\partial \xi}{\partial x} + \psi v \frac{\partial \xi}{\partial y} \right) \quad \text{at } z = \psi\xi(x, y, t). \tag{8}$$

(3) The solid-bottom boundary condition is defined as

$$w = -\varepsilon^2 \left(u \frac{\partial h}{\partial x} + v \frac{\partial h}{\partial y} \right) \quad at\ z = -h(x, y). \tag{9}$$

The continuity equation is integrated from the sea bottom to surface. Then, the following expression is obtained:

$$\varepsilon^2 \int_{-h}^{\psi\xi} \frac{\partial}{\partial x} u\, dz + \varepsilon^2 \int_{-h}^{\psi\xi} \frac{\partial}{\partial y} v\, dz = - \int_{-h}^{\psi\xi} \frac{\partial w}{\partial z} dz. \tag{10}$$

Vertically integrating each term of Eq. (10) using Leibniz's rule, and applying seabed boundary condition Eq. (9) and kinematic boundary conditions Eqs. (8) and (10) is written as

$$\frac{\partial}{\partial x} \int_{-h}^{z} u\, dz + \frac{\partial}{\partial y} \int_{-h}^{z} v\, dz + \frac{\partial \xi}{\partial t} = 0. \tag{11}$$

The momentum equations given in Eq. (4) is integrated over the depth using the boundary conditions in Eqs. (7)–(9), the following form is obtained (for simplification, p_{atm} is assumed as zero):

$$\frac{\partial}{\partial t} \int_{-h}^{\psi\xi} u\, dz + \psi \frac{\partial}{\partial x} \int_{-h}^{\psi\xi} u^2\, dz + \psi \frac{\partial}{\partial y} \int_{-h}^{\psi\xi} u\, v\, dz + \frac{\partial}{\partial x} \int_{-h}^{\psi\xi} p\, dz$$
$$- p(x, y, -h, t) \frac{\partial h}{\partial x} = 0. \tag{12}$$

Similarly, the integrated form of Eq. (5) is written as

$$\frac{\partial}{\partial t} \int_{-h}^{\psi\xi} v dz + \psi \frac{\partial}{\partial x} \int_{-h}^{\psi\xi} u\, v dz + \psi \frac{\partial}{\partial y} \int_{-h}^{\psi\xi} v^2\, dz + \frac{\partial}{\partial y} \int_{-h}^{\psi\xi} p\, dz$$
$$- p(x, y, -h, t) \frac{\partial h}{\partial y} = 0. \tag{13}$$

The pressure is calculated from the vertical-momentum equation given in Eq. (6) by integrating with respect to z and using the boundary conditions given in Eqs. (7) and (8) on the free surface.

$$p = \xi - \frac{z}{\psi} + \frac{\partial}{\partial t}\int_z^{\psi\xi} w dz + \psi\frac{\partial}{\partial x}\int_z^{\psi\xi} u\, w dz + \psi\frac{\partial}{\partial y}\int_z^{\psi\xi} v\, w dz - \frac{\psi}{\varepsilon^2}w^2. \tag{14}$$

Substituting all values of p, u, w in the depth and integrating continuity and horizontal-momentum equations (Eqs. (3)–(6)) using boundary conditions and after integrating the resulting equations, the following form of nonlinear BEs is obtained:

$$\frac{\partial \xi}{\partial t} + \nabla\cdot[(h+\psi\xi)\boldsymbol{u_\beta}] + \varepsilon^2\nabla\cdot\left(\frac{z_\beta^2}{2} - \frac{h^2}{6}\right)h\nabla(\nabla\cdot\boldsymbol{u_\beta}) + \varepsilon^2\nabla\cdot\left(z_\beta + \frac{h}{2}\right)h\nabla(\nabla\cdot(h\boldsymbol{u_\beta})) = 0, \tag{15}$$

$$\frac{\partial \boldsymbol{u_\beta}}{\partial t} + \nabla\xi + \psi(\boldsymbol{u_\beta}\cdot\nabla)\boldsymbol{u_\beta} + \varepsilon^2\frac{z_\beta^2}{2}\nabla\left(\nabla\cdot\frac{\partial \boldsymbol{u_\beta}}{\partial t}\right) + \varepsilon^2 z_\beta\nabla\left(\nabla\cdot\left(h\frac{\partial \boldsymbol{u_\beta}}{\partial t}\right)\right) = 0. \tag{16}$$

These are the nonlinear BEs which are utilized to analyze the nonlinear physical phenomena of water waves propagating horizontally over water of varying depth. Equations (15) and (16) are simplified to obtain the 1-D nonlinear BEs for constant water depth given as follows:

$$\frac{\partial \xi}{\partial t} + \frac{\partial u}{\partial x} + \psi\frac{\partial}{\partial x}(\xi u) + \varepsilon^2\left(\beta + \frac{1}{3}\right)\frac{\partial^3 u}{\partial^3 x} = 0, \tag{17}$$

$$\frac{\partial u}{\partial t} + \frac{\partial \xi}{\partial x} + \psi u\frac{\partial u}{\partial x} + \varepsilon^2\beta\frac{\partial^3 u}{\partial^2 x\partial t} = 0, \tag{18}$$

where $u = u(x,t)$ represents horizontal-velocity component at variable depth with $\beta = (z_\beta/h)^2/2 + z_\beta/h$. Here, z_β is the reference water depth, and h refers to constant water depth. Compared to the traditional form of the nonlinear BEs, Eqs. (15) and (16) are obtained by including an additional frequency dispersion term in the continuity equation. Equations (17) and (18) are combined to obtain the non-dimensional equation in one variable for the shallow-water wave transformation, and is given as

$$\frac{\partial^2 \xi}{\partial t^2} - \frac{\partial^2 \xi}{\partial x^2} - \psi \frac{\partial^2}{\partial t^2}\left(\xi^2\right) - \frac{\psi}{2}\frac{\partial^2}{\partial x^2}\left(\xi^2\right) + \varepsilon^2\left(2\beta + \frac{1}{3}\right)\frac{\partial^4 \xi}{\partial x^4} = 0. \tag{19}$$

The nonlinear BEs are capable of weakly spreading and nonlinear waves in shallow and intermediate water ($kh < 0.5$). The dimensional form of the nonlinear 1-D BEs is written as

$$\frac{\partial^2 \xi^*}{\partial t^2} - C^2 \frac{\partial^2 \xi^*}{\partial x^2} - \frac{\partial^2}{\partial t^2}\left(\xi^{*2}\right) - \frac{C^2}{2h}\frac{\partial^2}{\partial x^2}\left(\xi^{*2}\right) + C^2 h^2 \left(2\beta + \frac{1}{3}\right)\frac{\partial^4 \xi^*}{\partial x^4} = 0, \tag{20}$$

and the dimensional form of the 2-D nonlinear BEs is expressed as

$$\frac{\partial^2 \xi^*}{\partial t^2} - C^2\left(\frac{\partial^2 \xi^*}{\partial x^2} + \frac{\partial^2 \xi^*}{\partial y^2}\right) - \frac{\partial^2}{\partial t^2}\left(\xi^{*2}\right) - \frac{C^2}{2h}\left(\frac{\partial^2}{\partial x^2} + \frac{\partial^2}{\partial y^2}\right)\left(\xi^{*2}\right) + C^2 h^2\left(2\beta + \frac{1}{3}\right)\left(\frac{\partial^4 \xi^*}{\partial x^4} + \frac{\partial^4 \xi^*}{\partial x^2 \partial y^2} + \frac{\partial^4 \xi^*}{\partial y^4}\right) = 0, \tag{21}$$

where C is the non-dispersive phase speed for the shallow water.

3. Dispersion Characteristics

3.1. *Linear Analysis*

The comparison of the linear-dispersion relation based on present form of nonlinear BEs and the exact (i.e., Airy wave theory) dispersion relation of the BEs for different water depths is discussed. Equations (15) and (16) have been transformed to linearized forms given as

$$\frac{\partial \xi}{\partial t} + h\frac{\partial u}{\partial x} + \left(\beta + \frac{1}{3}\right) h^3 \frac{\partial^3 u}{\partial x^3} = 0, \tag{22}$$

$$\frac{\partial u}{\partial t} + g\frac{\partial \xi}{\partial x} + \beta h^2 \frac{\partial^3 u}{\partial x^2 \partial t} = 0. \tag{23}$$

Consider recurrence wave with frequency ω and wave number k given as

$$\xi(x,t) = a_0 \exp(i(kx - \omega t)), \quad u(x,t) = u_0 \exp(i(kx - \omega t)), \tag{24}$$

where a_0 and u_0 denotes the wave amplitude and velocity amplitude, respectively. Substituting Eq. (24) in Eqs. (22) and (23) and letting discriminant as zero to obtain the non-trivial solution of the dispersion relation given as

$$C_B^2 = \frac{w^2}{k^2} = gh\left[\frac{1 - \left(\beta + \frac{1}{3}\right)(kh)^2}{1 - \beta(kh)^2}\right], \tag{25}$$

where C_B is the phase speed of the present numerical model and the wave number is $k = 2\pi/L$. The exact linear-dispersion relation obtained from Airy wave theory is

$$C_{Airy}^2 = \frac{w^2}{k^2} = gh\left[\frac{\tanh(kh)}{kh}\right]. \tag{26}$$

Figure 1(a) shows that the normalized forms of phase speed and group velocity with Airy wave theory is displayed for different values of $\beta = \{0, -0.333, -0.39, -0.40, -0.50\}$. The abscissa is the relative water depth computed as a ratio of water depth h to corresponding deep-water wavelength $L = 2\pi g/w^2$. The deep-water limit depth corresponding to relative depth h/L is 0.5. If the velocity at the bottom $z_\beta = -h$ is selected, then $\beta = -0.50$. Alternatively, when the velocity at the mean water level is selected, then $\beta = 0$. The depth-averaged velocity of the standard form of the BEs corresponds to $\beta = -1/3$. The parameter β in BEs develops the linear-dispersion characteristics of the system. The concept of group velocity associated with wave energy is significant for the study of wave propagation and is

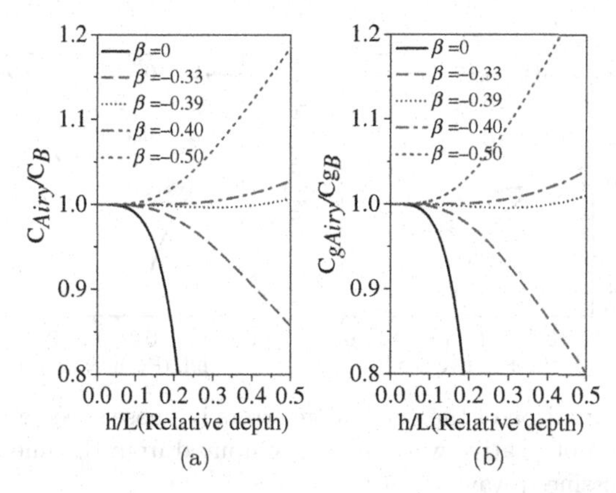

Figure 1. Normalized form of the phase speed and group velocity of present Boussinesq theory with Airy wave theory (C_B/C_{Airy}) and (Cg_B/Cg_{Airy}) for different values of $\beta = \{0, -0.33, -0.39, -0.40, -0.50\}$.

related with phase velocity as follows:

$$C_{gB} = \frac{d\omega}{dk} = C\left(1 - \frac{\frac{(kh)^2}{3}}{[1-\beta(k\ h)^2]\left[1-\left(\beta+\frac{1}{3}\right)(k\ h)^2\right]}\right). \tag{27}$$

The group velocities normalized with respect to Airy wave theory are plotted in Figure 1(b) for the different values of β, in which abscissa represents the relative water-depth function. For the intermediate relative water depth less than 0.3, the difference between the behavior of group velocities and phase speeds for the Boussinesq model and Airy wave theory are quite similar. The comparison of linear dispersion behavior of present model and linear-wave theory is shown in Figure 2. The percentage error graphs for the phase and group velocities are given, in which the error is computed from the expression

$$Percentage\ Error = \left(\frac{C_{Boussinesq} - C_{Airy}}{C_{Airy}}\right) \times 100. \tag{28}$$

Comparison of the phase speeds determined by the linear-wave theory and Boussinesq model is given in Figure 2(a). The percentage error is calculated by the formula in Eq. (28). The phase speed of nonlinear BEs deviates sharply from linear-wave theory, and the error exceeds 5% for a relative depth equal to 0.3. Similarly, Figure 2(b)

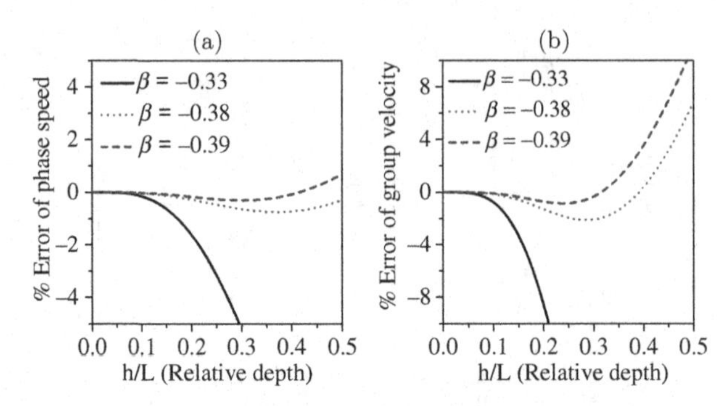

Figure 2. Comparison of the percentage errors in phase speed and group velocities as a function of relative water depth computed from the linear-wave theory and present Boussinesq wave theory.

shows the comparison of the group velocities determined by the linear-wave theory and Boussinesq model. The group velocity of nonlinear BEs deviates sharply from linear-wave theory, and the error is greater than 10% with relative depth greater than 0.2.

3.2. *Nonlinear Analysis*

Different frequency components interact at the harmonic of the primary-frequency waves to produce wave components that are defined for the propagation of surface waves. The nonlinear phenomenon of water waves is a combination of two wave trains with different frequencies. The Laplace equations show the significance of the forced harmonics for bidirectional, dichromatic waves. The weakly nonlinear BEs are capable of simulating the propagation of higher-order forced waves. Assume a wave train with two considerable amplitudes of recurrence waves with amplitudes a_1, a_2 and frequencies ω_1, ω_2. The surface elevation is defined as

$$\xi^{(1)}(x,t) = a_1 \cos(k_1 x - \omega_1 t) + a_2 \cos(k_2 x - \omega_2 t), \tag{29}$$

where k_1 and k_2 represent the wave numbers. The existence wave number appeases the first-order linearized form of the BEs. Therefore, the horizontal velocity is written as

$$u_\beta^1(x,t) = \frac{\omega_1}{k_1^1 h} a_1 \cos(k_1 x - \omega_1 t) + \frac{\omega_2}{k_2^1 h} a_2 \cos(k_2 x - \omega_2 t), \tag{30}$$

where $k_1^1 = k_1\big(1 - (\beta + \frac{1}{3})(k_1 h)^2\big)$ and $k_2^1 = k_2\big(1 - (\beta + \frac{1}{3})(k_2 h)^2\big)$.

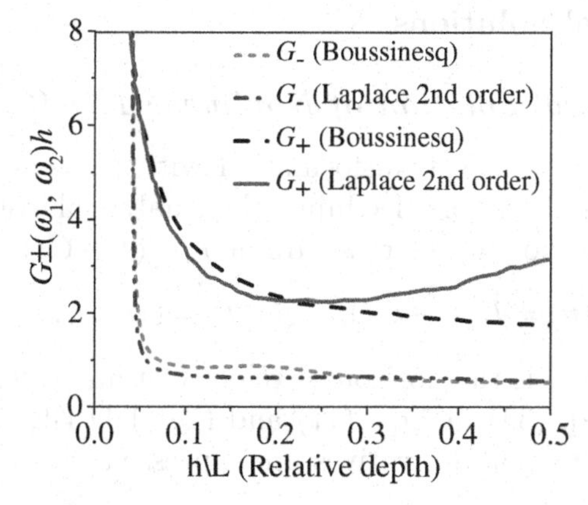

Figure 3. Comparison between the Boussinesq model and quadratic transfer function with second-order Laplace theories.

The equation for the forced waves that satisfy the BEs at second-order wave amplitude is written as

$$\frac{\partial \xi^2}{\partial t} + h\frac{\partial u_\beta^2}{\partial x} + \left(\beta + \frac{1}{3}\right) h^2 \frac{\partial^3 u_\beta^2}{\partial x^3} = -\xi^1 \frac{\partial u_\beta^1}{\partial x} - u_\beta^1 \frac{\partial \xi^1}{\partial x}, \tag{31}$$

$$\frac{\partial u_\beta^2}{\partial t} + g\frac{\partial \xi^2}{\partial x} + \beta h^2 \frac{\partial^3 u_\beta^2}{\partial x^2 \partial t} = -u_\beta^1 \frac{\partial u_\beta^1}{\partial t}. \tag{32}$$

The second-order wave is contained in a sub-harmonic at a different frequency $\omega_- = \omega_1 + \omega_2$ and higher harmonics at the sum frequencies $2\omega_1$, $2\omega_2$, and $\omega_+ = \omega_1 + \omega_2$. On solving the equations for amplitudes of the second-order surface wave height and velocity, the following explanation is obtained for quadratic transfer function:

$$G_\pm(\omega_1, \omega_2) = \frac{\omega_\pm k_\pm h G_1 \left[\omega_1 k_2^1 + \omega_2 k_1^1\right] + \omega_1 \omega_2 (k_\pm h)^2 G_2}{\left[\omega_\pm^2 G_1 - g k_\pm^2 h G_2 \left(2 k_1^1 k_2^1 h^3\right)\right]}, \tag{33}$$

where $G_1 = \left[1 - \beta (k_\pm h)^2\right]$ and $G_2 = \left[1 - \left(\beta + \frac{1}{3}\right)(k_\pm h)^2\right]$

Figure 3 shows the quadratic transfer function of the BEs compared to that of the second-order Laplace equation. The quadratic transfer functions were calculated for bidirectional waves with different angles of partitions between respective wave components. The magnitude of the down wave and second-harmonic set at the depth of the deep water are defined for the Boussinesq model. Therefore, it does not accurately simulate non-linear wave effects in deep water.

4. Analytical Solutions

4.1. *Analytical Solution of Nonlinear 1-D BEs*

Assume a solitary-wave transformation with phase speed C in the positive x–axis direction. Defining the analytical solution of BEs depending on a moving reference frame $\nu = (x - Ct)$ as

$$u(x,t) = U(\nu) = U(x - Ct), \quad \xi(x,t) = \xi(\nu) = \xi(x - Ct), \tag{34}$$

where $u = U(x,t)$ is the horizontal velocity at an arbitrary depth z_β. Substituting Eq. (34) in Eqs. (17) and (18) 1-D BEs and using the chain rule, the partial derivatives are expressed as

$$\frac{\partial u}{\partial x} = \frac{\partial U}{\partial \nu}\frac{\partial \nu}{\partial x} = 1.\frac{\partial U}{\partial \nu} = \frac{\partial U}{\partial \nu}, \quad \frac{\partial U}{\partial t} = \frac{\partial U}{\partial \nu}\frac{\partial \nu}{\partial t} = -C\frac{\partial U}{\partial \nu}. \tag{35}$$

Similarly,

$$\frac{\partial^3 u}{\partial x^3} = \frac{\partial^3 U}{\partial v^3}, \quad \frac{\partial \xi}{\partial t} = -C\frac{\partial \xi}{\partial \nu}, \quad \frac{\partial \xi}{\partial x} = \frac{\partial \xi}{\partial \nu}. \tag{36}$$

Then,

$$\varepsilon^2\left(\beta + \frac{1}{3}\right)\frac{\partial^3 U}{\partial \nu^3} + \psi\frac{\partial}{\partial \nu}(\xi U) + \frac{\partial U}{\partial \nu} - C\frac{\partial \xi}{\partial \nu} = 0, \tag{37}$$

$$-\varepsilon^2 \beta C\frac{\partial^3 U}{\partial \nu^3} + \psi U\frac{\partial U}{\partial \nu} + \frac{\partial U}{\partial \nu} - \frac{\partial \xi}{\partial \nu} = 0. \tag{38}$$

Integrating Eqs. (37) and (38) with zero-integral constants and eliminating, we get

$$\varepsilon^2(\beta + \frac{1}{3})\frac{\partial^2 U}{\partial \nu^2} + \psi(\xi U) + U - C\xi = 0, \tag{39}$$

$$\xi = U - \psi\frac{U^2}{2} - \varepsilon^2 \beta C\frac{\partial^2 U}{\partial \nu^2}. \tag{40}$$

Substitute Eq. (40) into Eq. (39), and on simplifying, we get

$$- \psi\varepsilon^2\beta C U\frac{\partial^2 U}{\partial \nu^2} + \left(\varepsilon^2\left(\beta + \frac{1}{3} - \varepsilon^2\beta C^2\right)\right)\frac{\partial^2 U}{\partial \nu^2}$$
$$- \psi^2\frac{U^3}{2} + \frac{3}{2}\psi C U^2 + (1 - C^2)U = 0. \tag{41}$$

Equation (41) is calculated using the first-integral method (FIM). Assume $U(\nu)$ is a solution of a first-order ODE:

$$\phi(U) = \left(\frac{\partial U}{\partial \nu}\right)^2, \quad \frac{\partial \phi(U)}{\partial U} = 2\frac{\partial U^2}{\partial \nu^2}. \tag{42}$$

Let us assume that the solution is of the form

$$U(\nu) = u_0\, sech^2(k\nu), \quad k > 0, \tag{43}$$

where u_0 is the velocity amplitude and k is the wave number. Substituting Eq. (43) in Eqs. (42) and (43), the equations are obtained as

$$\begin{aligned} &\frac{\partial U}{\partial \nu} = 2kU\, tanh(k\nu), \quad \left(\frac{\partial U}{\partial \nu}\right)^2 = 4k^2U^2\left(1 - sech^2(k\nu)\right), \\ &\frac{\partial^2 U}{\partial \nu^2} = AU - \frac{3B}{2}U^2, \quad A = 4k^2, \quad B = -\frac{4k^2}{u_0}. \end{aligned} \tag{44}$$

Substituting Eq. (44) in Eq. (41), we get the third-degree homogeneous polynomial in U, and to define the non-trivial solution U, setting all the coefficients of this polynomial to be zero:

$$-C(C + \varepsilon^2 A\beta C) + \varepsilon^2\left(\beta + \frac{1}{3}\right)A + 1 = 0, \tag{45}$$

$$\begin{aligned} &-\psi(C + \varepsilon^2 A\beta C) - C\left(\frac{3}{2}\varepsilon^2\beta CB - \frac{\psi}{2}\right) \\ &+\frac{3}{2}\varepsilon^2\left(\beta + \frac{1}{3}\right)B = 0, \end{aligned} \tag{46}$$

$$\psi\left(-\frac{1}{2}\psi + \frac{3}{2}\varepsilon^2 B\beta C\right) = 0. \tag{47}$$

Solving Eqs. (45)–(47) analytically, the results are obtained as

$$\begin{aligned} &A = -\frac{1}{2}\frac{(9\beta + 1)}{2\varepsilon^2\beta(3\beta + 1)}, \quad B = \pm\frac{\psi(3\beta - 1)}{\varepsilon^2(3\beta + 1)\sqrt{\beta\left(\beta + \frac{1}{3}\right)}}, \\ &C = \pm\frac{\left(\beta + \frac{1}{3}\right)}{\sqrt{\beta\left(\beta + \frac{1}{3}\right)}}. \end{aligned} \tag{48}$$

Substituting A and B values in Eq. (44), we get

$$k = \frac{\sqrt{6}}{12}\sqrt{\frac{(9\beta+1)}{-\beta\varepsilon^2\left(\beta+\frac{1}{3}\right)}}, \quad u_0 = \frac{\sqrt{3}(9\beta+1)}{2}\frac{\sqrt{\beta(3\beta-1)}}{\beta\psi(3\beta-1)}, \tag{49}$$

which yields Eq. (43). Using Eq. (43) in Eq. (38), the resulting equation is expressed as

$$\begin{aligned} &\xi(x,t) = \xi_0\, sech^2(k(x-Ct)), \quad \xi_0 = -\frac{1}{4}\frac{(9\beta+1)}{\psi\beta}, \\ &u(x,t) = u_0\, sech^2(k(x-Ct)). \end{aligned} \tag{50}$$

Figure 4 shows that the solitary wave attends the peak (trough or crest), and solitary waves have the same behavior. Here, we have defined the surface elevation of solitary waves at constant slopes for the 1-D nonlinear BEs model and propagate elliptical holes at constant water depth. Analytical solutions for multiple elevations for surface elevation are compared with other models with represent the importance of higher-order nonlinear terms in 1-D BEs model.

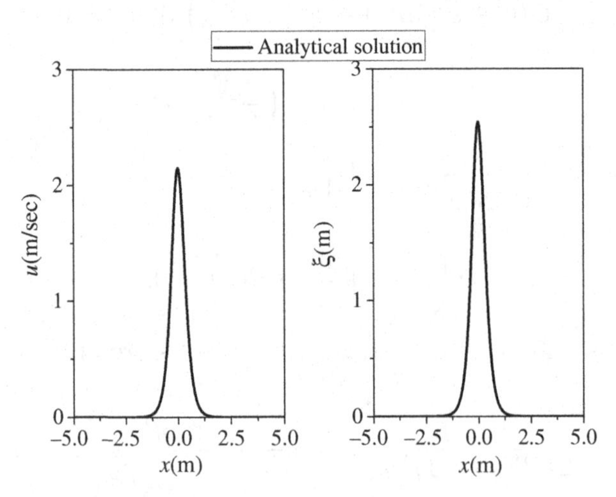

Figure 4. Analytical solution of the 1-D BEs for defined surface elevation and water-wave velocity at water depth 0.56 m and $\beta = -0.39$ at $t = 20$ s.

4.2. *Analytical Solution of Nonlinear 2-D BEs*

Nwogu's nonlinear 2-D BEs are rewritten as

$$\frac{\partial \xi}{\partial t} + \frac{\partial}{\partial x}\left[(h+\xi)\,u\right] + \frac{\partial}{\partial y}\left[(h+\xi)\,v\right] + (a_1+a_2)\,h^3 \left[\frac{\partial^3 u}{\partial x^3} + \frac{\partial^3 v}{\partial x^2 \partial y} + \frac{\partial^3 u}{\partial x \partial y^2} + \frac{\partial^3 v}{\partial y^3}\right] = 0, \quad (51)$$

$$\frac{\partial u}{\partial t} + g\frac{\partial \xi}{\partial x} + u\frac{\partial u}{\partial x} + v\frac{\partial u}{\partial y} + (b_1+b_2)\,h^2 \times \left[\frac{\partial^3 u}{\partial x^2 \partial t} + \frac{\partial^3 v}{\partial x \partial y \partial t}\right] = 0, \quad (52)$$

$$\frac{\partial v}{\partial t} + g\frac{\partial \xi}{\partial y} + u\frac{\partial v}{\partial x} + v\frac{\partial v}{\partial y} + (b_1+b_2)\,h^2 \times \left[\frac{\partial^3 u}{\partial x \partial y \partial t} + \frac{\partial^3 v}{\partial y^2 \partial t}\right] = 0, \quad (53)$$

where a_1, a_2, b_1, and b_2 are constants given as $a_1 = \left(\frac{\beta^2}{2} - \frac{1}{6}\right)$, $a_2 = \left(\beta + \frac{1}{2}\right)$, $b_1 = \frac{\beta^2}{2}$, and $b_2 = \beta$. Here, $\beta = \frac{z_\beta}{h}$. Equation (51) is the continuity equation, and Eqs. (52) and (53) are momentum equations in x and y directions. Alternative to water-wave velocity u, we use the velocity potential ϕ (such that $u = \phi$) as the dependent variable. Then,

$$\frac{\partial \phi}{\partial t} + h\left(\frac{\partial^2 \phi}{\partial x^2} + \frac{\partial^2 \phi}{\partial y^2}\right) + \frac{\partial}{\partial x}\left(\phi\frac{\partial \phi}{\partial x}\right) + \frac{\partial}{\partial y}\left(\phi\frac{\partial \phi}{\partial y}\right) + (a_1+a_2)\,h^3\left(\frac{\partial^4 \phi}{\partial x^4} + \frac{\partial^4 \phi}{\partial x^2 \partial y^2} + \frac{\partial^4 \phi}{\partial y^4}\right) = 0, \quad (54)$$

$$\frac{\partial^2 \phi}{\partial x \partial t} + g\frac{\partial \xi}{\partial x} + \frac{\partial \phi}{\partial x}\frac{\partial^2 \phi}{\partial x^2} + \frac{\partial \phi}{\partial y}\frac{\partial^2 \phi}{\partial x \partial y} + (b_1+b_2)\,h^2 \left(\frac{\partial^4 \phi}{\partial x \partial y^2 \partial t} + \frac{\partial^4 \phi}{\partial x^3 \partial t}\right) = 0, \quad (55)$$

$$\frac{\partial^2\phi}{\partial y\partial t}+g\frac{\partial\xi}{\partial y}+\frac{\partial\phi}{\partial x}\frac{\partial^2\phi}{\partial x\partial y}+\frac{\partial\phi}{\partial y}\frac{\partial^2\phi}{\partial y^2}+(b_1+b_2)\,h^2$$
$$\left(\frac{\partial^4\phi}{\partial x^2\partial y\partial t}+\frac{\partial^4\phi}{\partial y^3\partial t}\right)=0. \quad (56)$$

From Eqs. (52) and (53),

$$\frac{\partial\xi}{\partial x}=-\frac{1}{g}\left[\frac{\partial^2\phi}{\partial x\partial t}+\frac{\partial\phi}{\partial x}\frac{\partial^2\phi}{\partial x^2}+\frac{\partial\phi}{\partial y}\frac{\partial^2\phi}{\partial x\partial y}+(b_1+b_2)\,h^2\right.$$
$$\left.\left(\frac{\partial^4\phi}{\partial x\partial y^2\partial t}+\frac{\partial^4\phi}{\partial x^3\partial t}\right)\right], \quad (57)$$

$$\frac{\partial\xi}{\partial y}=-\frac{1}{g}\left[\frac{\partial^2\phi}{\partial y\partial t}+\frac{\partial\phi}{\partial x}\frac{\partial^2\phi}{\partial x\partial y}+\frac{\partial\phi}{\partial y}\frac{\partial^2\phi}{\partial y^2}+(b_1+b_2)\,h^2\right.$$
$$\left.\left(\frac{\partial^4\phi}{\partial x^2\partial y\partial t}+\frac{\partial^4\phi}{\partial y^3\partial t}\right)\right]. \quad (58)$$

Integrating Eqs. (57) and (58) with respect to x and y, respectively,

$$\xi=-\frac{1}{g}\left[\frac{\partial\phi}{\partial t}+\frac{1}{2}\left(\left(\frac{\partial\phi}{\partial x}\right)^2+\left(\frac{\partial\phi}{\partial y}\right)^2\right)+(b_1+b_2)\,h^2\right.$$
$$\left.\left(\frac{\partial^3\phi}{\partial x^2\partial t}+\frac{\partial^3\phi}{\partial y^2\partial t}\right)\right]. \quad (59)$$

Differentiating Eq. (59) with respect to t,

$$\frac{\partial\xi}{\partial t}=-\frac{1}{g}\left[\frac{\partial^2\phi}{\partial t^2}+\frac{1}{2}\frac{\partial}{\partial t}\left(\left(\frac{\partial\phi}{\partial x}\right)^2+\left(\frac{\partial\phi}{\partial y}\right)^2\right)\right.$$
$$\left.+(b_1+b_2)\,h^2\left(\frac{\partial^3\phi}{\partial x^2\partial t^2}+\frac{\partial^3\phi}{\partial y^2\partial t^2}\right)\right]. \quad (60)$$

Substituting all values in the continuity equation in Eq. (51),

$$
\begin{aligned}
&-\frac{\partial^2\phi}{\partial t^2}-\frac{1}{2}\frac{\partial}{\partial t}\left(\left(\frac{\partial\phi}{\partial x}\right)^2-\left(\frac{\partial\phi}{\partial y}\right)^2\right)-(b_1+b_2)\,h^2\\
&\times\left(\frac{\partial^3\phi}{\partial x^2\partial t^2}+\frac{\partial^3\phi}{\partial y^2\partial t^2}\right)+\left(hg-\left(\frac{\partial\phi}{\partial t}+\frac{1}{2}\left(\left(\frac{\partial\phi}{\partial x}\right)^2+\left(\frac{\partial\phi}{\partial y}\right)^2\right)\right.\right.\\
&\left.\left.+(b_1+b_2)\,h^2\left(\frac{\partial^3\phi}{\partial x^2\partial t}+\frac{\partial^3\phi}{\partial y^2\partial t}\right)\right)\right)\left(\frac{\partial^2\phi}{\partial x^2}+\frac{\partial^2\phi}{\partial y^2}\right)\\
&-\frac{\partial\phi}{\partial x}\left(\frac{\partial^2\phi}{\partial x\partial t}+\frac{\partial\phi}{\partial x}\frac{\partial^2\phi}{\partial x^2}+\frac{\partial\phi}{\partial y}\frac{\partial^2\phi}{\partial x\partial y}\right)-h^2\,(b_1+b_2)\,\frac{\partial\phi}{\partial x}\\
&\times\left(\frac{\partial^4\phi}{\partial x\partial y^2\partial t}+\frac{\partial^4\phi}{\partial x^3\partial t}\right)-\frac{\partial\phi}{\partial y}\left(\frac{\partial^2\phi}{\partial y\partial t}+\frac{\partial\phi}{\partial x}\frac{\partial^2\phi}{\partial x\partial y}+\frac{\partial\phi}{\partial y}\frac{\partial^2\phi}{\partial y^2}\right)\\
&-h^2\,(b_1+b_2)\,\frac{\partial\phi}{\partial y}\left(\frac{\partial^4\phi}{\partial x^2\partial y\partial t}+\frac{\partial^4\phi}{\partial y^3\partial t}\right)+(a_1+a_2)\,h^3\\
&\times\left(\frac{\partial^2\phi}{\partial x^2}+\frac{\partial^2\phi}{\partial y^2}\right)^2=0. \qquad (61)
\end{aligned}
$$

Assume a solitary wave propagating with a phase velocity of C in the positive direction of the x axis. This is simply to find solitary-wave solutions, which depend only on the reference value ζ. Assuming

$$\zeta = kx + hy - Ct, \quad \phi(\zeta) = \phi(kx + hy - Ct). \qquad (62)$$

Then, differentiating Eq. (61) with a new variable ζ with respect to x, y, and t, we have

$$\frac{\partial\phi}{\partial x} = k\phi', \quad \frac{\partial\phi}{\partial y} = h\phi', \frac{\partial\phi}{\partial t} = -C\phi', \qquad (63)$$

where the prime denotes the differentiation with respect to the new variable ζ.

$$
\begin{aligned}
&(hg(h^2+k^2)-C^2)\phi''-\frac{1}{2}(h^2+k^2)\phi'^2\\
&+2C(h^2+k^2)\phi'\phi''-\frac{1}{2}(h^2+k^2)^2\phi'^2\phi''\\
&-k^2(h^2+k^2)\phi'^2\phi''-(b_1+b_2)(h^2+k^2)h^2C^2\phi''''
\end{aligned}
$$

$$+ (b_1 + b_2)(h^2 + k^2)^2 h^2 C\phi'''\phi'' + (b_1 + b_2)h^2 k^3 C\phi''''\phi'$$
$$- (h^2 + k^2)^2 h^2 \phi''\phi' + h^4 C(b_1 + b_2)(h^2 + k^2)^2 \phi''''\phi'$$
$$+ gh^3(a_1 + a_2)(h^2 + k^2)^2 \phi'''' = 0. \tag{64}$$

The truncated terms which are neglected from the above equation are defined as

$$o(\phi^2, \phi^2\phi'') = -\frac{1}{2}(h^2 + k^2)\phi'^2 - \frac{1}{2}(h^2 + k^2)^2 \phi'^2\phi''$$
$$- k^2(h^2 + k^2)\phi'^2\phi''. \tag{65}$$

Then, integrating Eq. (63) with variable ζ, we get

$$(hg(h^2 + k^2) - C^2)\phi' + C(h^2 + k^2)\phi'^2$$
$$- (b_1 + b_2)(h^2 + k^2)h^2 C^2 \phi''' + (b_1 + b_2)(h^2 + k^2)^2 h^2 C \frac{\phi''^2}{2}$$
$$+ (b_1 + b_2)h^2 k^3 C \frac{\phi'^3}{3} - (h^2 + k^2)^2 h^2 \frac{\phi'^2}{2}$$
$$+ (b_1 + b_2)(h^2 + k^2)^2 h^4 C \frac{\phi'^3}{3}$$
$$+ gh^3(a_1 + a_2)(h^2 + k^2)^2 \phi''' = G_1. \tag{66}$$

Again, integrating the equation which is obtained by integration and multiplying $2\phi''$ and then simplifying the result, we have

$$(hg(h^2 + k^2) - C^2)(\phi')^2 + \left[C - \frac{(h^2 + k^2)}{2}h^2\right](h^2 + k^2)\frac{2}{3}(\phi')^3 - h^2$$
$$(h^2 + k^2)\left[(b_1 + b_2)C^2 + gh(a_1 + a_2)(h^2 + k^2)\right](\phi'')^2 + (b_1 + b_2)h^2 C$$
$$\left[k^3 + h^2(h^2 + k^2)\right]\frac{(\phi')^2}{12} + (b_1 + b_2)(h^2 + k^2)^2 \frac{(\phi'')^3}{3} = 2G_1\phi' + G_2. \tag{67}$$

Now, G_1 and G_2 are zero for solitary waves, where $\phi' = \phi'' = \phi''' = \phi'''' = 0$ as $|\zeta| \to \infty$. Then, assuming the solution form ϕ

is expressed as

$$\phi' = A\, sech^2\left(B\zeta\right), \quad u(x,y,t) = A\, sech^2\left(B(kx+hy-Ct)\right), \tag{68}$$

where

$$\begin{aligned} A &= \left(\frac{gh(h^2+k^2)}{C^2}\right)^{\frac{1}{2}}, \\ B &= \left[\frac{C^2 - gh(h^2+k^2)k^4}{4(h^2+k^2)h^2\left[(b_1+b_2)C^2+(a_1+a_2)gh(h^2+k^2)\right]}\right]^{\frac{1}{2}}, \end{aligned} \tag{69}$$

$$\begin{aligned} \xi(x,y,t) &= A_1 sech^2\left(B(kx+hy-Ct)\right) + A_2 sech^4\left(B(kx+hy-Ct)\right), \\ A_1 &= \left[\frac{C^2 - gh(h^2+k^2)}{3(h^2+k^2)\left[(b_1+b_2)C^2-(a_1+a_2)gh(h^2+k^2)\right]}\right] h, \\ A_2 &= -\frac{\left[C^2-gh(h^2+k^2)\right]\ \left[(b_1+b_2)gh^2+2(a_1+a_2)C^2(h^2+k^2)\right]}{2ghC^2(h^2+k^2)\left[(b_1+b_2)gh^2-(a_1+a_2)C^2(h^2+k^2)\right]} h. \end{aligned} \tag{70}$$

Hence, these are the solutions of nonlinear 2-D BEs for water-wave propagation. Figure 5 shows the solution of the 2-D BEs for the water-wave propagation over a closed rectangular region. We evaluate the solitary-wave propagation for 2-D BEs model in terms of water-wave velocity and surface-wave height. For the 2-D BEs model, the surface height has a larger wave amplitude in addition to the water-wave velocity. This model is defined to determine the water-wave propagation of monochromatic waves above submerged shoals, and the model is capable of propagating shallow-water ripples for a coastal region.

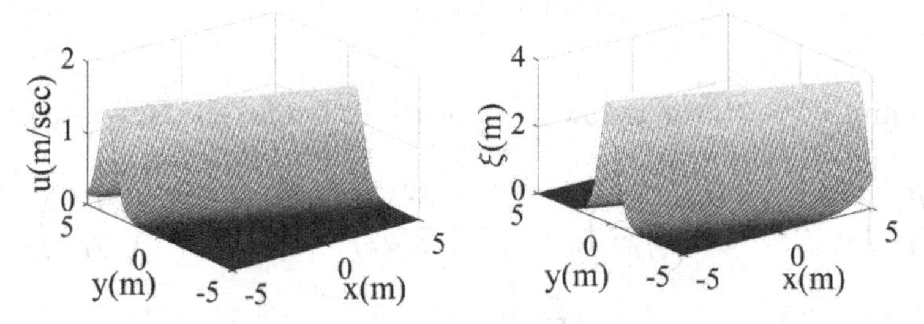

Figure 5. Analytical solution of the 2-D BEs defined for surface elevation and water-wave velocity at water depth 0.56 m and $\beta = -0.39$.

5. Numerical Solutions of 1-D and 2-D BEs

The Crank–Nicolson discretization with the predictor–corrector procedure is employed to obtain the numerical solution of the system given in Eqs. (20) and (21). Finite-difference scheme approximates the partial derivatives for time (t) variable and central-difference scheme approximates the derivatives corresponding to space variables. Here, Δx is the space-step size, and Δt refers to the time-step size. The solution procedure consist of three stages. At time step $t = n\Delta t$, the value of variable $t = (n + (1/2)\Delta t$ is predicted using the given values at $t = n\Delta t$. Then, the approximated values at time $t = (n + (1/2)\Delta t$ are utilized in the corrector stage to determine the values at time $t = (n + 1)\Delta t$. Finally, the estimated values attained at time $t = (n+1)\Delta t$ are considered as initial approximations for the iterative scheme. The value of ξ at the grid point $(x, t) = (i\Delta x,\ n\ \Delta t)$ is denoted by ξ_i^n, where $i = 1, 2, 3, \ldots, (N + 1)$ and $n = 1, 2, 3, \ldots$.

5.1. *Discretization Form of Nonlinear 1-D BEs*

An iterative Crank–Nicolson scheme for discretizing the weakly nonlinear and dispersive BEs is used. In addition, this scheme is accurate for the second order of rationalizing the predicate stage to obtain forecast values at new time steps. Finite-difference scheme is used for partial derivations in time and space variables.

The dimensional form of Eq. (20) is rewritten as

$$\frac{\partial^2 \xi}{\partial t^2} - C^2 \frac{\partial^2 \xi}{\partial x^2} - \frac{\partial^2}{\partial t^2}\left(\xi^2\right) - \frac{C^2}{2h}\frac{\partial^2}{\partial x^2}\left(\xi^2\right) + C^2 h^2 \left(2\beta + \frac{1}{3}\right)\frac{\partial^4 \xi}{\partial x^4} = 0. \tag{71}$$

Assuming $F = \xi^2$, the following approximation for the BEs is defined:

$$\left(\frac{\partial^2 \xi}{\partial t^2}\right)^{(n+1)} - C^2\left(\frac{\partial^2 \xi}{\partial x^2}\right)^{(n+1)} - \left(\frac{\partial^2 F}{\partial t^2}\right)^{(n+1)} + C^2 h^2\left(\frac{\partial^4 \xi}{\partial x^4}\right)^{(n+1)} \left(2\beta + \frac{1}{3}\right) - \frac{C^2}{2\,h}\left(\frac{\partial^2 F}{\partial x^2}\right)^{(n)} = 0. \tag{72}$$

The discretized form of nonlinear BEs in Eq. (72) is given as follows:

$$\frac{\xi_i^{n+1} - 2\xi_i^n + \xi_i^{n-1}}{\Delta t^2}$$
$$- \frac{C^2}{2}\left(\frac{\xi_{i+1}^n - 2\xi_i^n + \xi_{i-1}^n}{\Delta x^2} + \frac{\xi_{i+1}^{n+1} - 2\xi_i^{n+1} + \xi_{i-1}^{n+1}}{\Delta x^2}\right)$$
$$- \left(\frac{F_i^{n+1} - 2F_i^n + F_i^{n-1}}{\Delta t^2}\right) - \frac{C^2}{2h}\left(\frac{F_{i+1}^n - 2F_i^n + F_{i-1}^n}{\Delta x^2}\right)$$
$$+ \left(2\beta + \frac{1}{3}\right)\frac{C^2h^2}{2}\left(\frac{\xi_{i+2}^{n+1} - 4\xi_{i+1}^{n+1} + 6\xi_i^{n+1} - 4\xi_{i-1}^{n+1} + \xi_{i-2}^{n+1}}{\Delta x^4}\right)$$
$$+ \left(\frac{\xi_{i+2}^n - 4\xi_{i+1}^n + 6\xi_i^n - 4\xi_{i-1}^n + \xi_{i-2}^n}{\Delta x^4}\right)\left(2\beta + \frac{1}{3}\right)\frac{C^2h^2}{2} = 0. \tag{73}$$

Equation (73) is rewritten as follows:

$$r_1\xi_{i+2}^{n+1} - (r + 4r_1)\,\xi_{i+1}^{n+1} + (1 - 2r + 6r_1)\,\xi_i^{n+1} - (r + 4r_1)$$
$$\xi_{i-1}^{n+1} - r_1\xi_{i-2}^{n+1} = -r_1\,\xi_{i+2}^n + (r + 4r_1)\,\xi_{i+1}^n + (2 - 2r - 6r_1)\,\xi_i^n$$
$$+ (r + 4r_1)\,\xi_{i-1}^n + r_1\,\xi_{i-2}^n + r_2\left(F_{i+1}^n - 2F_i^n + F_{i-1}^n\right) - \xi_i^{n-1}$$
$$+ \left(F_{i+1}^n - 2F_i^n + F_{i-1}^n\right).$$

The resulting system of equations is expressed as

$$r_1\,\xi_{i+2}^{n+1} - (r + 4\,r_1)\,\xi_{i+1}^{n+1} + (1 - 2\,r + 6\,r_1)\,\xi_i^{n+1} - (r + 4\,r_1)$$
$$\xi_{i-1}^{n+1} - r_1\,\xi_{i-2}^{n+1} = S_i. \tag{74}$$

The right-hand side of Eq. (74) is S_i, which is given as $S_i = -r_1\,\xi_{i+2}^n + (r + 4\,r_1)\,\xi_{i+1}^n + (2 - 2r - 6\,r_1)\,\xi_i^n + (r + 4\,r_1)\,\xi_{i-1}^n + r_1\,\xi_{i-2}^n + r_2\,\left(F_{i+1}^n - 2F_i^n + F_{i-1}^n\right) - \xi_i^{n-1} + \left(F_{i+1}^n - 2F_i^n + F_{i-1}^n\right)$, where $i = 1, 2, 3, \ldots, N-1$, n is a positive integer, and the coefficient values are

$$r = \frac{C^2\Delta t^2}{2\,\Delta x^2}, \quad r_1 = \frac{C^2h^2\Delta t^2}{2\Delta x^4}\left(2\beta + \frac{1}{3}\right),$$
$$\text{and} \quad r_2 = \frac{1}{2h}\left(\frac{C^2\Delta t^2}{\Delta x^2}\right).$$

The above discretization simplifies to give the following iterative form: $R\xi_i^{n+1} = S_i$. The discretization form of nonlinear BEs is represented in the form of penta-diagonal system of equations. At each iteration, the Gauss elimination with partial-pivoting scheme is utilized to obtain the solution of the system in matrix form.

5.2. *Discretization Form of Nonlinear 2-D BEs*

The numerical discretization is defined by a staggered-grid system, and the meshes are defined as $i = 1, 2, \ldots, m$ in the x direction and $j = 1, 2, \ldots, n$ in the y direction. The mesh sizes in the x and y directions are represented by Δx and Δy, respectively and the time step by Δt. The nonlinear 2-D BEs (Eq. (21)) are discretized using iterative Crank–Nicholson scheme. Assuming $F = \xi^2$, the following approximation for the BEs is defined:

$$\left(\frac{\partial^2 \xi}{\partial t^2}\right)^{(n+1)} - C^2\left(\frac{\partial^2 \xi}{\partial x^2} + \frac{\partial^2 \xi}{\partial y^2}\right)^{(n+1)} - \left(\frac{\partial^2 F}{\partial t^2}\right)^{(n)} - \left(\frac{\partial^2 F}{\partial x^2} + \frac{\partial^2 F}{\partial y^2}\right)^{(n)} \frac{C^2}{2h} + C^2 h^2 \left(2\beta + \frac{1}{3}\right) \times \left(\frac{\partial^4 \xi}{\partial x^4} + \frac{\partial^4 \xi}{\partial x^2 \partial y^2} + \frac{\partial^4 \xi}{\partial y^4}\right)^{(n+1)} = 0. \tag{75}$$

The discretized form of nonlinear BEs in Eq. (75) is given as follows:

$$\frac{\xi_{i,j}^{n+1} - 2\xi_{i,j}^{n} + \xi_{i,j}^{n-1}}{\Delta t^2} - \left(\frac{\xi_{i+1,j}^{n} - 2\xi_{i,j}^{n} + \xi_{i-1,j}^{n} + \xi_{i+1,j}^{n+1} - 2\xi_{i,j}^{n+1} + \xi_{i-1,j}^{n+1}}{\Delta x^2}\right)\frac{C^2}{2} - \left(\frac{\xi_{i,j+1}^{n} - 2\xi_{i,j}^{n} + \xi_{i,j-1}^{n}}{\Delta y^2} + \frac{\xi_{i,j+1}^{n+1} - 2\xi_{i,j}^{n+1} + \xi_{i,j-1}^{n+1}}{\Delta y^2}\right)\frac{C^2}{2} - \frac{C^2}{2h}\left(\frac{F_{i+1,j}^{n} - 2F_{i,j}^{n} + F_{i-1,j}^{n}}{\Delta x^2} + \frac{F_{i,j+1}^{n} - 2F_{i,j}^{n} + F_{i,j-1}^{n}}{\Delta y^2}\right)$$

$$+\left(2\beta+\frac{1}{3}\right)\frac{C^2h^2}{2}(a+a_2+b_1+b_2+c_1+c_2)$$

$$-\left(\frac{F_i^{n+1}-2F_i^n+F_i^{n-1}}{\Delta t^2}\right)=0, \tag{76}$$

where

$$a_1=\left(\frac{\xi_{i+2,j}^{n+1}-4\xi_{i+1,j}^{n+1}+6\xi_{i,j}^{n+1}-4\xi_{i-1,j}^{n+1}+\xi_{i-2,j}^{n+1}}{\Delta x^4}\right),$$

$$a_2=\left(\frac{\xi_{i+2,j}^{n}-4\xi_{i+1,j}^{n}+6\xi_{i,j}^{n}-4\xi_{i-1,j}^{n}+\xi_{i-2,j}^{n}}{\Delta x^4}\right),$$

$$b_1=\left(\frac{\xi_{i,j+2}^{n+1}-4\xi_{i,j+1}^{n+1}+6\xi_{i,j}^{n+1}-4\xi_{i,j-1}^{n+1}+\xi_{i,j-2}^{n+1}}{\Delta y^4}\right),$$

$$b_2=\left(\frac{\xi_{i,j+2}^{n}-4\xi_{i,j+1}^{n}+6\xi_{i,j}^{n}-4\xi_{i,j-1}^{n}+\xi_{i,j-2}^{n}}{\Delta y^4}\right),$$

$$c_1=\left(\frac{\xi_{i+1,j+1}^{n}-2\xi_{i,j+1}^{n}+\xi_{i-1,j+1}^{n}-2\left(\xi_{i+1,j}^{n}-2\xi_{i,j}^{n}+\xi_{i-1,j}^{n}\right)}{\Delta x^2\Delta y^2}\right.$$

$$\left.+\frac{\xi_{i+1,j-1}^{n}-2\xi_{i,j-1}^{n}+\xi_{i-1,j-1}^{n}}{\Delta x^2\Delta y^2}\right),$$

$$c_2=\left(\frac{\xi_{i+1,j+1}^{n+1}-2\xi_{i,j+1}^{n+1}+\xi_{i-1,j+1}^{n+1}-2\left(\xi_{i+1,j}^{n+1}-2\xi_{i,j}^{n+1}+\xi_{i-1,j}^{n+1}\right)}{\Delta x^2\Delta y^2}\right.$$

$$\left.+\frac{\xi_{i+1,j-1}^{n+1}-2\xi_{i,j-1}^{n+1}+\xi_{i-1,j-1}^{n+1}}{\Delta x^2\Delta y^2}\right).$$

Similarly, the discretization form is defined as a system of equations in iterative form $R_1\xi^{n+1}=S_1\xi^n$, where R_1 and S_1 are both penta-diagonal matrices with respect to x and y directions, respectively. Similarly, Eq. (5.18) is solved by finite-difference method using Crank–Nicolson scheme. Using this scheme, we have to find tridiagonal systems of linear equations with constant coefficients. This means that only the right-hand side of such a scheme changes at each time level. Hence, the solution of nonlinear BEs is obtained by using TDMA (Thomas algorithm).

5.3. *1-D Boundary Conditions*

Consider the mathematical domain as $[x_0, x_l]$ with grid size Δx such that $x_j = j\Delta x$, where $j = 0, 1, 2, \ldots, N$. Here, $1, 2, 3, \ldots, N-1$ are the interior points of the domain, and $j = 0$ and $j = N$ are the boundary points. The boundary condition at $x = 0$ (incident boundary) is given as

$$\xi(0,t) = \frac{(kh)}{\omega}\left(1 - \left(\beta + \frac{1}{3}\right)(kh)^2\right)u(0,t). \tag{77}$$

The second-order partial derivatives for the velocity $\xi_{xx}(0,t)$ are also specified at the incident boundary. The perfectly absorbing boundary condition is given as

$$\frac{\partial \xi}{\partial t} + C_0 \frac{\partial \xi}{\partial x} = 0, \tag{78}$$

where $C_0 = \omega/k$ is the phase speed of water waves. The first-order partial derivative for the wave height at the outgoing absorbing boundary is given as

$$\frac{\partial \xi}{\partial x}|_{N,j} = \frac{3\xi_{N,j} - 4\xi_{N-1,j} + \xi_{N-2,j}}{2\Delta x}. \tag{79}$$

If the wave reflection is significant at the incident boundary, then the incident boundary condition is modified to include fully absorbing boundary condition as follows:

$$\frac{\partial \xi}{\partial t} - C_0 \frac{\partial \xi}{\partial x} - 2\frac{\partial \xi_0}{\partial t} = 0, \tag{80}$$

where $\partial \xi_0 / \partial t$ is the time-derivative term of the incident-wave elevation.

5.4. *2-D Boundary Conditions*

The nonlinear 2-D BEs model was defined for the wave propagation in a rectangular basin surrounded by four vertical sides. If the continuity equation contains a source term, the waves can be generated internally, following [7], at a constant water depth:

$$\frac{\partial \xi}{\partial t} + h\Delta \cdot u + \left(\beta + \frac{1}{3}\right) h^3 \Delta^2 (\Delta \cdot u) = f(x, y, t). \tag{81}$$

In a constant water depth of h, a plane wave is generated with a small amplitude a_0 and water-wave angular frequency ω. The angle between the propagation direction of the wave and the x axis is θ. The source function $f(x, y, t)$ has two parts:

$$f(x, y, t) = g(x)s(y, t), \tag{82}$$

where $g(x)$ is a Gaussian hump function and $s(y, t)$ is the magnitude of the source function with input time series, which are defined as

$$g(x) = \exp\left[-\beta_0 (x - x_c)^2\right], \quad s(y, t) = D\, sin(k_y y - \omega t),$$

where β_0 is the source function of the shape coefficient and x_c is the center coordinate of the source in the x direction. $k_y = k\, sin(\theta)$ is the y-direction wave number, and k is the linear wave number. D is the magnitude of the source function:

$$D = \frac{2a_0 cos(\theta)\left(\omega^2 - \gamma g k^4 h^3\right)}{\omega k I\left(1 - \beta(kh)^2\right)}, \tag{83}$$

where β and γ are coefficients, and I is the integral that is given as

$$I = \sqrt{\frac{\pi}{\beta_0}} \exp(-k_x^2/4\beta_0), \tag{84}$$

where $k_x = kcos(\theta)$ is the wave number in the x direction. The shape coefficient β_0 is defined as any number. The larger the value of β_0, the narrower the source function. In this chapter, β_0 is given by $\beta_0 = \frac{80}{\delta^2 L^2}$, where L is the wavelength and δ is of order one. The wave will be reflected completely when it reaches a solid wall. For a general reflective boundary with an outward normal vector $\vec{\boldsymbol{n}}$, we would anticipate that the boundary conditions would be

$$u \cdot \vec{\boldsymbol{n}} = 0, \quad \xi \cdot \vec{\boldsymbol{n}} = 0. \tag{85}$$

The initial conditions for the Gaussian hump of water are given as

$$\xi(x, y, t = 0) = H_0 \exp\left[-\beta\left((x - x_c)^2 + (y - y_c)^2\right)\right], \tag{86}$$

$$u(x, y, t = 0) = 0, \quad v(x, y, t = 0) = 0, \tag{87}$$

where H_0 is the initial height of the hump, β is shape coefficient, which controls the width of the hump, and (x_c, y_c) is the coordinate at the center of the domain.

6. Results

6.1. *Validation Results*

To estimate the effect of nonlinear wave on the bathymetry with time t, the surface elevation of weakly nonlinear and nonlinear Boussinesq equations are compared with linear Boussinesq equation for $T = 1.0\,\text{s}$ at the following three different locations: (a) $x = 10\,\text{m}$, (b) $x = 15\,\text{m}$, and (c) $x = 20\,\text{m}$, obtained from $t = 40$ s to $t = 45\,\text{s}$, with grid size $\Delta x = 0.025\,\text{m}$ and $\Delta t = 0.02\,\text{s}$ time step. Figure 6 shows that the wave-surface elevations obtained from weakly nonlinear and nonlinear Boussinesq equations have small deviations when compared to that obtained from linear Boussinesq model. The wave elevation of long irregular waves along positive x axis is defined as additional number of wave coefficients as

$$\xi(x,t) = \sum_{j=1}^{N} A_j sin(w_j t - k_j x + \phi_j), \tag{88}$$

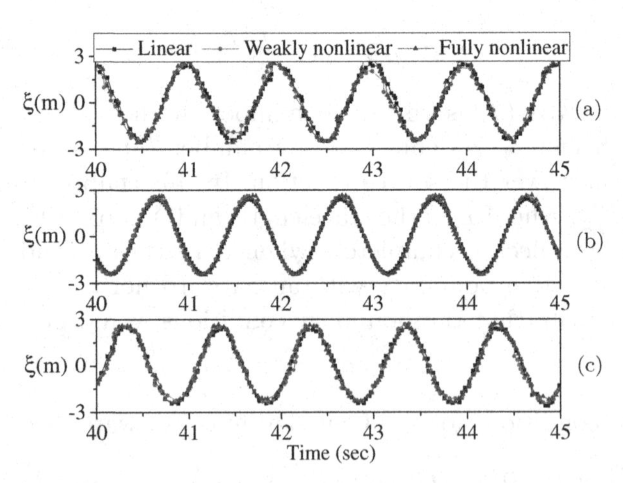

Figure 6. Comparison of the time series with surface-elevation profile for constant depth $h = 0.25\,\text{m}$ and time interval for spatial variable x: (a) $x = 10\,\text{m}$, (b) $x = 15\,\text{m}$, and (c) $x = 20\,\text{m}$.

where A_j is the mean of wave amplitude, ω_j is the angular frequency of water waves, k_j is the wave number, and ϕ_j is a random phase angle of wave components and it is equally distributed between 0 and 2π with time. In Figure 7, the surface elevation with irregular waves is evaluated with time series of 10–30 s at the entrance boundary of depth of water of $h = 0.56\,\text{m}$ with wave period $T = 0.35\,\text{s}$. Here, the time interval of the grid size is $\Delta t = 0.02\,\text{s}$, and the space interval of grid size $\Delta x = 0.025\,\text{m}$, and the free parameter related to water depth is $\beta = -0.39$. The relative water depth at the incident boundary is 1 m. The Boussinesq model is now used to investigate the oscillation of irregular waves in deep to intermediate and shallow waters, defined as wave-group characteristics in irregular waves.

In this study, the surface wave evolution is defined in the closed basin of size $L_x = 10\,\text{m} \times L_y = 10\,\text{m}$ with constant water depth $h_0 = 0.56\,\text{m}$. The linear analytical solution is obtained by first extending the domain by the method of images in both x and y directions, leading to the waveform that is periodic over $2L_x$ and $2L_y$ and even about any image of the reflecting domain walls. The initial maximum elevation H_0 is taken to be 0.45 m, with a corresponding height to depth ratio 0.1. The model is run for 100 s using a grid size of $\Delta x = 0.15\,\text{m}$ and time step of $\Delta t = 0.05\,\text{s}$. Figure 8 shows the effects of large initial nonlinearity on the calculated solution. If we take an

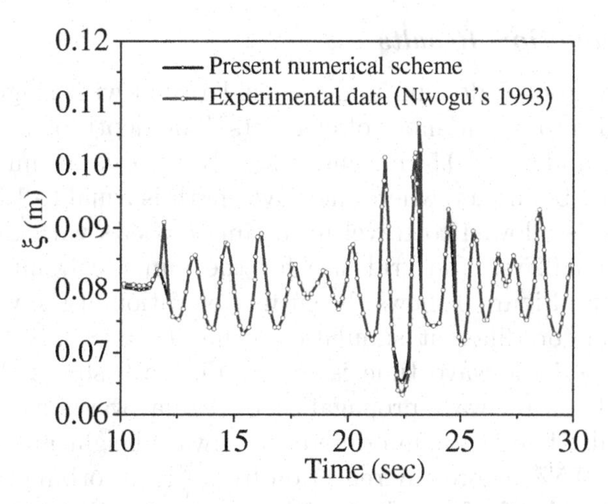

Figure 7. Validation of time-series plot of the surface elevation for irregular waves with the experimental data at depth 0.24 m and $\beta = -0.39$.

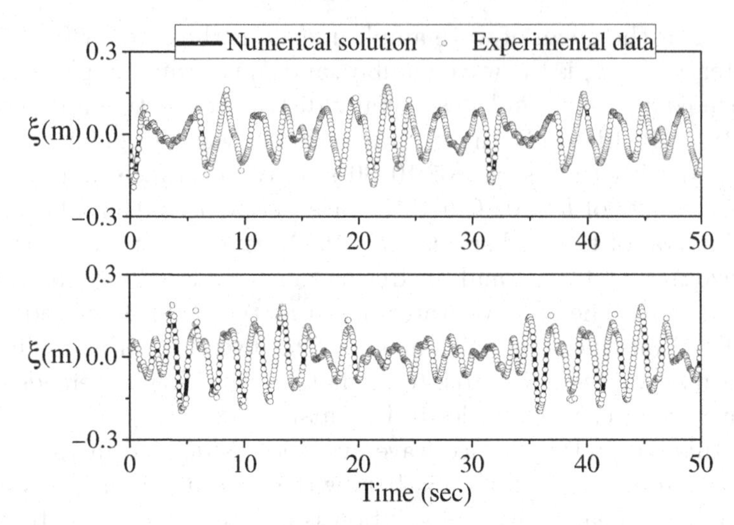

Figure 8. Comparison of solution of surface elevation for initial Gaussian elevation in rectangular region with nonlinear analytical solutions.

initial shape, the height is $H_0 = 0.45$ m. This corresponds to the initial height at water depth of 1, which is outside the normal range of validity of the model. This relative depth ratio drops dramatically as the initial hill of water expands.

6.2. *Simulation Results*

The computational domain under consideration with length 15 m is partitioned into 450 number of elements. The depth of water at the incident boundary is 0.56 m, and a steady wave is an immersed at the outgoing boundary, where the wavelength is equal to 1.56 m. The water wave is allowed to travel until time $t = 35$ s with time difference $\Delta t = 0.025$ s. The grid size for the numerical simulations is $\Delta x = 0.02$ m. Figure 9 shows the surface elevation of the water-wave propagation for different simulation times $t = (8.5, 17, 25.5, 34)$ s. Here, the periodic-wave time is 0.85 s. Figure 9 shows the spatial profile of the water-wave propagation of regular waves with different time periods. The maximum difference of wave height in the channel is less than 0.5% before any reflection from the absorbing range. The BEs are capable of effectively modeling the generation of deep-water waves in constant-depth waters.

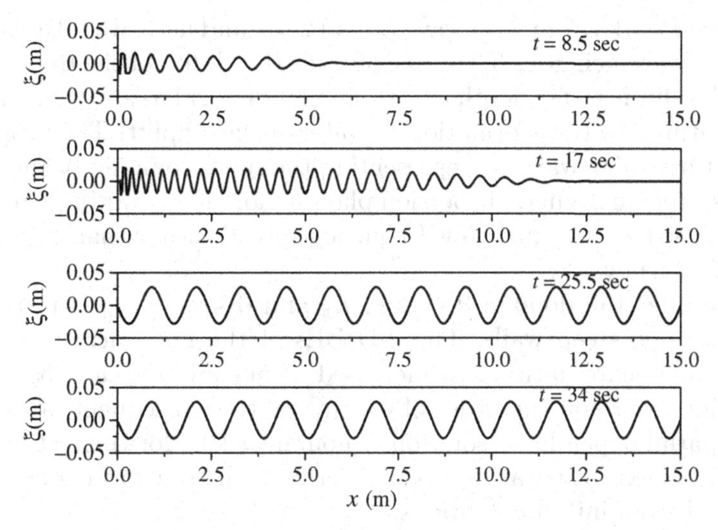

Figure 9. Plot of the surface elevation for regular waves with evaluated different simulation times $t = 8.5, 17, 25.5$, and 34 s at depth $h = 0.56\,\text{m}$.

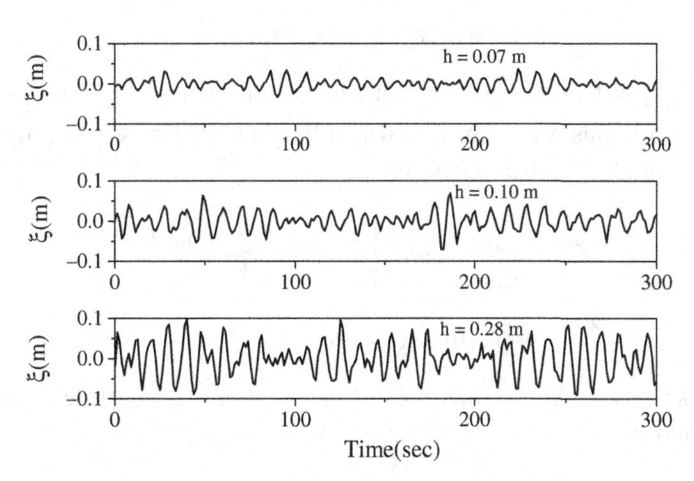

Figure 10. Surface elevation of irregular-waves propagation with respect to time for different depths 0.07, 0.10, and 0.28 m.

Figure 10 shows the surface elevation of irregular-waves propagation with respect to time for different depths 0.07 m, 0.10 m, and 0.28 m. The numerical model is used with grid spacing $\Delta x = 0.04\,\text{m}$, and the time-step size is $\Delta t = 0.02\,\text{s}$. The wavelength of the water waves is $L = 1.127$ m with wave period 2.5 s, and here, we take the significant wave height as 3.3 m. We observe the ability of the Boussinesq model to limit water depth and to simulate irregular-wave generation illustrating the transformation of water-surface uplift. The propagation of irregular waves is representing the more complex behavior of shallow regions, where dispersion plays a more important role of non-criticality, i.e., extremely low frequency spreads faster than extremely high frequency.

Consider the domain $0 < x < L_x$ and $0 < y < L_y$ is restricted by reflective steep walls. The 2-D BEs of the numerical model, to study monochromatic-wave increased frequency over a shoal takes on values more on the order of 0.2 during the subsequent processes. The spatial dependence solution of nonlinear BEs for the rectangular domain is expressed as two cosine series with transform coefficients resolved from initial conditions:

$$\xi_{nm} = \frac{1}{(1+\delta_{n0})(1+\delta_{m0})} \int_{-L_x}^{L_x} \int_{-L_y}^{L_y} \xi_0(x,y) \times cos(n\lambda x) cos(m\lambda y) dx\, dy, \tag{89}$$

where δ_{nm} is the Kronecker delta function with $\lambda = \frac{\pi}{L_x} = \frac{\pi}{L_y}$. Each (n, m) mode has a matching natural frequency (Boussinesq and Airy wave theory), which is given as

$$C_B^2 = \frac{w_{nm}^2}{k_{nm}^2} = gh_0 \left[\frac{1 - (\beta + \frac{1}{3})(k_{nm}h_0)^2}{1 - \beta(k_{nm}h_0)^2} \right], \tag{90}$$
$$\omega_{nm}^2 = gk_{nm} tanh(k_{nm}h_0),$$

where $k_{nm}^2 = (n\lambda)^2 + (m\lambda)^2 = (\frac{\pi}{L_x})^2(n^2 + m^2)$. The linear solution is defined as

$$\xi(x, y, t) = \sum_{n=1}^{\infty} \sum_{m=1}^{\infty} \xi_{nm} e^{-i\omega_{nm}t} cos(n\lambda x) cos(m\lambda y). \tag{91}$$

This is the real part of the linear solution. For water-wave propagation, the initial surface elevation is defined in Gaussian hump as

$$\xi(x, y) = H_0 \exp\left[-2\left((x - 3.75)^2 + (y - 3.75)^2\right)\right], \tag{92}$$

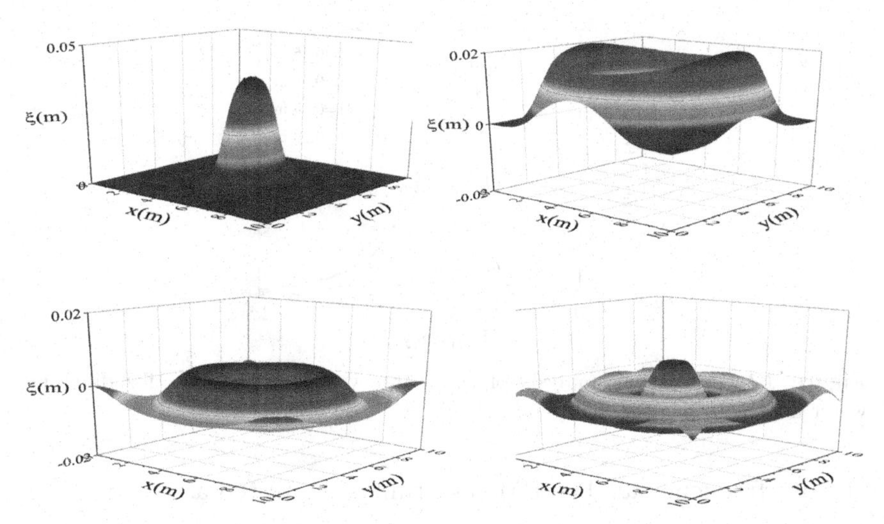

Figure 11. Profiles of surface elevation of water-waves propagation for different times $t = 10$, 20, 30, and 40 s with water depth $h = 0.56$ m.

where x and y are defined with respect to the origin in the left and bottom corner of the rectangular tank, and the initial surface elevation is symmetric about the center ($x = 3.75\,\text{m}, y = 3.75\,\text{m}$). Figure 11 shows the profiles of surface elevation at times $t = 0$, 20, 30, 40 s for illustrative purposes. The model was run for 100 s, using a grid size of $\Delta x = 0.025\,\text{m}$ and time step of $\Delta t = 0.06\,\text{s}$. The linear sloshing mode is computed for time $t = 2.64\,\text{s}$ in regular-frequency domain. The computational results are less than one for the total volume percent error. The results are computed using joint boundary conditions, which were more exact than normal velocity conditions with zero, and the results obtained using governing equations alone. The axisymmetry of the evolving wave about the origin of the Gaussian hump is distinct from the contour plot of the surface elevation in $t = 1\,\text{s}$.

We define the Pierson–Moskowitz (PM) spectrum to be too restricted for fully established seas of North Atlantic ocean, where there is local balance between transfer from the wind. The PM spectrum is defined for fully established sea in deep water with one spectrum parameter. Significant wave height (H_s) is related to the

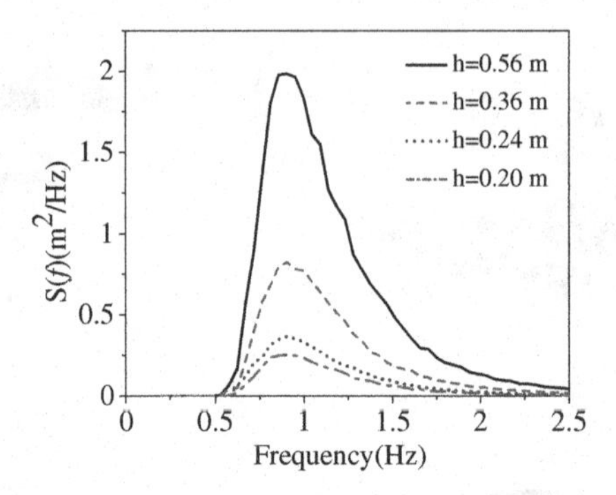

Figure 12. Spectrum analysis of the water-waves propagation for different water depths 0.56, 0.36, 0.24, and 0.20 m.

angular frequency ω. The PM spectrum is expressed as

$$S(f) = \frac{8.10 \times 10^{-3} g^2}{2\pi^4 \omega^5} \cdot \exp\left[-0.032 \frac{gH_s^2}{2\pi\omega^2}\right], \tag{93}$$

where H_s is the significant wave height. Figure 12 shows that the spectrum analysis of the water-waves propagation for different water depths 0.56, 0.36, 0.24, and 0.20 m with $H_s = 0.09$ m. This spectrum is developed by offshore areas fully developed which are generated local winds this is one parameter spectrum. The spectral analysis is defined by averaging over a frequency band of width 0.04 Hz. As the wave train shoals, there is an increment with the low and high frequency wave energies due to amplification. So, the magnitude of PM spectrum decreases whenever the water depths decrease.

7. Conclusion

The nonlinear 1-D and 2-D BEs are used to develop an efficient numerical model to reproduce the wave revolution in coastal regions. The dispersion characteristics of nonlinear BEs for water-wave propagation are defined, which defines the applicability of BEs model for shallow to deep waters. The numerical solution of BEs (linear, weakly nonlinear, nonlinear) is obtained by using fourth-order Finite-Difference Method (FDM) using predictor–corrector procedure. For 1-D and 2-D flows with constant water depth, the stability ranges are defined for the Boussinesq model, which are evaluated by the Eigen values with amplification matrix, and the scheme will be stable if the Courant number is less than 1. Linear stability analysis is defined by von Neumann's method, which is to increment the ability and efficiency of the present numerical model, and the analytical solutions of 1-D and 2-D BEs are defined by first-integral method.

This chapter presents a 2-D horizontal numerical flow solution for modeling the propagation of waves in the coastal zone, for intermediate to shallow waters. Spectrum analysis is also defined to analyze the distribution of energy for water-wave propagation. The present numerical models have been used to analyze the variation in surface elevations for regular and irregular waves with respect to different time intervals, and time-series analysis is performed at different depths for regular and irregular waves of different surface elevations. The 2-D BE model is defined for highly nonlinear and spreading waves. In 2-D models, higher-derivative terms are created by numerical problems as well as uncertainties in boundary conditions. The present model is used to analyze the wave-oscillation effect at the coastal boundary, and further, the realistic harbor implementation has been used for the Boussinesq model.

References

[1] Peregrine, D. H. (1967). Long waves on a beach. *Journal of Fluid Mechanics*, 27, 815–827.

[2] Madsen, P. A., Murray, R., and Sorensen, O. R. (1991). A new form of the Boussinesq equations with improved linear dispersion characteristics. *Coastal Engineering*, 15, 371–388.

[3] Madsen, P. A., Murray, R., and Sorensen, O. R. (1992). A new form of the Boussinesq equations with improved linear dispersion characteristics. Part-2 A slowly varying bathymetry. *Coastal Engineering*, 18, 183–204.

[4] Beji, S. and Nadaoka, K. (1996). A formal derivation and numerical modeling of the improved Boussinesq equations for varying depth. *Ocean Engineering*, 23, 619–704.

[5] Nwogu, O. G. (1993). Alternative form of Boussinesq equations for near shore wave propagation. *ASCE Journal of Waterway Port, Coastal Ocean Engineering*, 119, 618–638.

[6] Kennedy, B., Chen, Q., Kirby, J. T., and Dalrymple, R. A. (2001). Boussinesq-type equations with improved nonlinear performance. *Wave Motion*, 33, 225–243.

[7] Wei, G. and Kirby, J. T. (1995). Time dependent numerical code for extended Boussinesq equations. *ASCE Journal of Waterway Port, Coastal Ocean Engineering*, 121, 251–261.

[8] Nwogu, O. G. (1997). Numerical prediction of breaking waves and currents with a Boussinesq model. In: *Coastal Engineering*, 25th International Conference on Coastal Engineering, September 2–6, 1996. Orlando, FL, pp. 4807–4820.

[9] Schäffer, H. A. and Madsen, P. A. (1998). Discussion of "A formal derivation and numerical modeling of the improved Boussinesq equations for varying depth". *Ocean Engineering*, 25(6), 497–500.

[10] Gobbi, M. F. and Kirby, J. T. (1999). Wave evolution over submerged sills: Tests of a high order Boussinesq model. *Coastal Engineering*, 37(1), 57–96.

[11] Bona, J. L. and Chen, M. (1998). A Boussinesq system for two-way propagation of nonlinear dispersive waves. *Physica D*, 116(1–2), 191–224.

[12] Walkley, M. (1999). A numerical method for extended Boussinesq shallow-water wave equations. PhD Thesis, School of Computer Studies, University of Leeds, UK.

[13] Kennedy, A. B., Chen, Q., Kirby, J. T., and Dalrymple, R. A. (2000). Boussinesq modeling of wave transformation, breaking, and runup. I: 1D. *ASCE Journal of Waterway, Port, Coastal, and Ocean Engineering*, 126(1), 39–47.

[14] Wei, G., Kirby, J. T., Grilli, S. T., and Subramanya, R. (1995). A fully nonlinear Boussinesq model for surface waves: Part I. Highly nonlinear unsteady waves. *Journal of Fluid Mechanics*, 294, 71–92.

[15] Gobbi, M. F., Kirby, J. T., and Wei, G. (2000). A fully nonlinear Boussinesq model for surface waves. Part 2. Extension to O(kh)4. *Journal of Fluid Mechanics*, 405, 181–210.

[16] Chen, Q., Madsen, P. A., Schäffer H. A., and Basco D. R. (2000). Wave-current interaction based on an enhanced Boussinesq approach. *Coastal Engineering*, 33, 11–39.

[17] Karambas, T. V. and Memo, C. D. (2009). Boussinesq model for weakly nonlinear fully dispersive water waves. *ASCE Journal of Waterway Port, Coastal and Ocean Engineering*, 135(5), 187–199.

[18] Lin, P. and Man, C. (2007). A staggered — Grid numerical algorithm for extended Boussinesq equations. *Applied Mathematical Modeling*, 31, 349–368.

[19] Chuanjian, M. (2003). Development of a two dimensional wave model based on the extended Boussinesq equations. PhD Thesis.

[20] Madsen, P. A., Bingham, H. B., and Liu, H. A. (2002). New Boussinesq method for fully nonlinear waves from shallow water to deep water. *Journal of Fluid Mechanics*, 462, 1–30.

[21] Kumar, P. and Gulshan. (2017). Extreme wave-induced oscillation in Paradip Port under the resonance conditions. *Pure and Applied Geophysics*, 174 (12), 1–16.
[22] Kumar, P. and Gulshan. (2018). Theoretical analysis of extreme wave oscillation in Paradip Port using a 3–D boundary element method. *Ocean Engineering*, 164, 13–22.
[23] Kumar, P., Rajni, and Rupali. (2018). Wave induced oscillation in an irregular domain by using hybrid finite element model wave induced oscillation in an irregular domain by using hybrid finite element model. *Journal of Physics: Conference Series*, 1039, 2018 8th International Conference on Applied Physics and Mathematics (ICAPM 2018), 27–29 January 2018, Phuket, Thailand.
[24] Lee, J. J. (1971). Wave-induced oscillations in harbour of arbitrary geometry. *Journal of Fluid Mechanics*, 45(2), 375–394.
[25] Wei, G. and Kirby, J. T. (1998). Simulation of water waves by Boussinesq models. *Journal of Waterway, Port, Coastal and Ocean Engineering*, 130(1), 1728.
[26] Kumar, H., Malik, A., Gautam, M. S., and Chand, F. (2017). Dynamics of shallow water waves with various Boussinesq equations. *Acta Phisyca Polonica A*, 131(2), 275–282.
[27] Mohapatra, S. C. and Soares, C. G. (2015). Comparing solutions of the coupled Boussinesq equations in shallow water. In: Carlos Guedes Soares, T.A. Santos (eds.), Maritime Technology and Engineering. Taylor Francis: London, pp. 947–954.
[28] Woo, S. B. and Liu, P. L. (2004). Finite-element model for modified Boussinesq equations II : Applications to nonlinear harbor oscillations. *Journal of Waterway, Port, Coastal and Ocean Engineering*, 130(1), 17–28.
[29] Gao, J., Ji, C., Gaidai, O., and Liu, Y. (2016). Numerical study of infra gravity waves amplication during harbor resonance. *Ocean Engineering*, 116, 90–100.
[30] Wang, G., Sun, Z. B., Gao, J. L., and Ma, X. Z. (2017). Numerical study of edge waves using extended Boussinesq equations. *Water Science and Engineering*, 10(4), 295–302.
[31] Lynett, P. J., Wu, T. R., and Liu, P. L. F. (2002). Modeling wave runup with depth-integrated equations. *Coastal Engineering*, 46, 89–107.
[32] Nwogu, O. and Demirbilek, Z. (2010). Infragravity wave motions and run-up over shallow fringing reefs. *Journal of Waterway, Port, Coastal and Ocean Engineering*, 136, 295–305.
[33] Hai, X. J., Chang-gen, L. I. U., and Jian hua, T. A. O. (2015). An extended form of Boussinesq-type equations for nonlinear water waves. *Journal of Hydrodynamics*, 27(5), 696–707.

[34] Schäffer, H. A., Madsen, P. A., and Deigaard, R. (1993). A Boussinesq model for waves breaking in shallow water, *Coastal Engineering*, 20, 185–202.

[35] Bonneton, P., Barthelemy, E., Chazel, F., Cienfuegos, R., Lannes, D., Marche, F., and Tissier, M. (2011a). Recent advances in Serre-Green Naghdi modelling for wave transformation, breaking, and runup. *European Journal of Mechanics-B/Fluids*, 30, 589–597.

[36] Kaihatu, J. M. and Kirby, J. T. (1998). Two-dimensional parabolic modeling of extended Boussinesq equations. *ASCE Journal of Waterway, Port, Coastal, and Ocean Engineering*, 124(2), 57–67.

[37] Gobbi, M. F., Kirby, J. T., and Wei, G. (2000). A fully nonlinear Boussinesq model for surface waves. Part 2. Extension to O(kh)4. *Journal of Fluid Mechanics*, 405, 181–210.

[38] Fang, K. Z., Zhang, Z., and Zhong, B. (2014). Modelling of 2–D extended Boussinesq equations using a hybrid numerical scheme. *Journal of Hydrodynamics*, 26(2), 187–198.

[39] Zhang, T., Lin, Z. H., Huang, G. Y., Fan, C. M., and Li, P. W. (2020). Solving Boussinesq equations with a mesh less finite difference method. *Ocean Engineering*, 15, 106957.

[40] Madsen, P. A. and Fuhrman, D. R. (2020). Trough instabilities in Boussinesq formulations for water waves. *Journal of Fluid Mechanics*, 889, A38.

[41] Gao, J., Ma, X., Zang, J., Dong, G., Ma, X., Zhu, Y., and Zhou, L. (2020). Numerical investigation of harbor oscillations induced by focused transient wave groups. *Coastal Engineering*, 15, 103670.

https://doi.org/10.1142/9789811245367_0004

Chapter 4

Differential Quadrature-based Advanced Numerical Scheme for Simulation of Solitary Motion of Shallow Water Waves

T. K. Pal[*,¶], D. Datta[†,‡,§,‖], and R. K. Bajpai[*,**]

[*] *Scientist, Technology Development Division, Bhabha Atomic Research Centre, Mumbai, India*

[†] *Former Scientist, Bhabha Atomic Research Centre, Mumbai, India*

[‡] *Adjunct Professor, Department of Mathematics, SRM Institute of Science & Technology, Chennai, India*

[§] *Visiting Professor, Department of Mechanical Engineering, GLA University, Mathura, UP, India*

[¶] *tkpal@barc.gov.in*

[‖] *dbbrt_datta@yahoo.com, debabrata.datta@gla.ac.in*

[**] *rkbajpai@barc.gov.in*

Abstract

Solitary wave motion, which arises in various fields of science and engineering such as hydrodynamics, quantum mechanics, plasma physics, and particle physics, is a nonlinear and weakly dispersive phenomenon generally modeled by a third-order nonlinear partial differential equation called the Korteweg–de Vries (KdV) equation. Our aim is to obtain a numerical solution of the governing KdV equation for shallow water waves using a robust and efficient numerical scheme called differential quadrature method (DQM), which has potential applications in solving higher-order differential equations, generally occurring in solid mechanics. One of the most promising advantages of DQM over standard numerical schemes, such as finite difference, finite element, and finite volume methods, is that it provides a very accurate and stable solution for a partial differential equation (PDE) using very few grid points. In fact, the solution becomes unstable if a large number of grid points are used in contrast to that of the standard numerical schemes, where solution becomes more accurate when spacing between grid points is reduced. This unique property of DQM makes it a fast algorithm for solving PDEs. Standard polynomial-based DQM with unequally spaced Gauss–Lobatto–Chebyshev points are used here for the simulation studies. The suggested numerical scheme is used for solving test problems involving a single solitary wave as well as interactions of two and three solitary waves. To check the accuracy of the newly applied method, the error norms, L_2 and L_∞, have been calculated. A fuzzy differential quadrature has been developed to solve the fuzzy KdV equation, which is formulated by considering the parameters associated with the nonlinear and dispersive terms of the KdV equation as fuzzy variables. The advantage of fuzzy differential quadrature is that we can reach the classical DQM solution from fuzzy DQM.

Keywords: solitary wave, shallow water wave, differential quadrature, fast algorithm, error norms, fuzzy differential quadrature

1. Introduction

We encounter motion of waves on the surface of a large body of water, caused by wind or tides at the beach, and the waves created by throwing a stone or by raindrops in water bodies, such as ponds, lakes, and rivers. Though all these water waves seem to be similar, they are all different types of water waves [1, 2]. In this chapter, we focus on shallow water waves, in which the depth of the water is much smaller than the wavelength of the disturbance of the free surface [3–5]. The ratio of the depth to the wave length of a shallow water wave is less than approximately 1/20. In shallow water bodies, waves are strongly affected by bottom depth, and high-amplitude waves distort because crests travel faster than troughs to form a profile with a steep rise and a slow fall [3]. Our emphasis will be especially on solitary wave motion of shallow water waves [6–8]. A solitary wave, as discovered experimentally by Scott Russell in 1844, is a localized wave that propagates with constant velocity and shape along one spatial direction only. The envelope of the wave has one global peak and decays far away from the peak. Russell observed the wave traveling along a channel of water while maintaining its original shape, and he called it a "great wave of translation" [9]. A special form of solitary wave is a soliton which has the additional property of retaining its permanent structure even after interacting with another soliton [6, 7]. For example, two solitons propagating in opposite directions effectively pass through each other without breaking. Solitary wave phenomenon arises in various fields of science and engineering, such as hydrodynamics, fiber optics, quantum mechanics, plasma physics, particle physics [10–17]. From physics point of view, a solitary wave is basically a nonlinear and weakly dispersive phenomenon, which is governed by the cancellation of nonlinear and dispersive effects in the medium [18].

A mathematical equation, famously known as the Korteweg–de Vries (KdV) equation, of propagation of long waves of small but finite amplitude in dispersive media was derived by Diederik Johannes Korteweg together with his Ph.D. student, Gustav de Vries in 1895 [19]. The KdV equation is used as a generic model for the study of weakly nonlinear long waves, incorporating leading-order nonlinearity and dispersion. Exact periodic solution of the KdV equation is known as cnoidal wave, and a special form of the solution in the limit of infinite wavelength represents a solitary wave [20, 21]. The KdV equation has been widely applied in those fields of science and engineering where solitary wave phenomenon occurs [10–17]. Researchers have derived analytical as well as numerical solutions of the KdV equation for particular problems [22–30]. In this chapter, a numerical solution of the KdV equation is presented using a robust and efficient numerical technique called differential quadrature method (DQM), which is relatively new in the field of numerical techniques for solving differential equations [31–33].

DQM is conceptually simple, and its implementation is quite straightforward, even for solving higher-order differential equations [31, 34]. One of the most promising advantages of DQM over standard numerical schemes, such as finite difference, finite element, and finite volume methods, is that it provides highly accurate and stable solution of ordinary and partial differential equations having globally smooth solutions using very few grid points. In fact, the solution becomes unstable if large number of grid points are used in contrast to that of the standard numerical schemes, where the solution becomes more accurate when spacing between grid points is reduced. This unique property of DQM makes it a fast algorithm for solving PDEs. So far, DQM has been successfully applied to boundary-value problems in various fields of science and engineering, such as structural mechanics and transport processes [31, 34–39]. DQM has also been used for solving differential equations with imprecise model parameters [40].

The chapter is structured as follows. In Section 2, the mathematical model of solitary wave in the form of the KdV equation is briefly introduced. In Section 3, the DQM based numerical framework of the KdV equation is formulated, in which the basics of traditional DQM with Lagrange interpolation polynomials as basis functions are discussed, and then, discretization of the KdV equation is carried out using the formulas of DQM. The discrete KdV equation is numerically solved for four benchmark problems and the accuracy of the results are estimated in Section 4.

2. Mathematical Model of Solitary Wave

When nonlinear and dispersive effects compensate with each other, a wave moving through a dissipative media, such as shallow water bodies, retains its original velocity and shape. A mathematical model of solitary wave works on this principle. Korteweg and de Vries incorporated effects of surface tension as a dispersive term in their model equation for shallow water waves, which is famously celebrated as the KdV equation [19]. They were able to derive closed-form periodic solutions of the KdV equation, which they named cnoidal waves. In the limit of infinite wavelength, the solution shows a single hump of positive elevation, which represents a solitary wave, as discovered experimentally by Scott Russell [9]. The KdV equation, in its simplest form, can be written as

$$\frac{\partial U(x,t)}{\partial t} + \varepsilon U(x,t)\frac{\partial U(x,t)}{\partial x} + \mu\frac{\partial^3 U(x)}{\partial x^3} = 0, \tag{1}$$

where ε and μ are constant parameters. The above equation is a nonlinear and weakly dispersive equation, where the nonlinearity is due to the second term and the dispersion is represented by the last term. The nonlinearity term tends to stiffen the wave form, while the dispersive term spreads the wave out. A delicate equilibrium between these two terms results in the formation of a soliton. In this chapter, Eq. (1), with the following initial and Dirichlet-type boundary conditions, is numerically solved using Lagrange polynomial-based DQM.

$$U(x, t = 0) = f(x),$$
$$U(x = x_L, t) = \alpha_1, \quad U(x = x_R, t) = \alpha_2,$$

where x_L and x_R are the left and right boundary positions, respectively, and α_1 and α_2 are the values of the solution at the respective boundary locations.

3. Numerical Framework

In this section, an overview of the mathematical background of DQM for solving general PDEs and discretization of the governing KdV equation using the DQM formulation is presented.

3.1. *Differential Quadrature Method*

Differential quadrature (DQ) is a numerical technique for approximating derivatives of a sufficiently smooth function, originally developed by Bellman and Casti using a simple analogy with integral quadrature, where integration of a definite integral is approximated as the weighted sum of functional values in the whole integral domain [41, 42]. The basic philosophy behind DQM is the concept that the partial derivative of a field variable at any discrete point in the computational domain can be approximated by a weighted linear sum of the values of the field variable along the line that passes through that point, which is parallel to the coordinate direction of the derivative. The mathematical formulation of this philosophy can be framed as

$$f_x^m(x_i) = \left.\frac{\partial^m f(x)}{\partial x^m}\right|_{x=x_i} = \sum_{k=1}^{N} A_{ik}^{(m)} f(x_k), \quad i = 1, 2, \ldots, N; \qquad (2)$$
$$m = 1, 2, \ldots, N - 1,$$

where x_i are the discrete points of the coordinate system, m is the order of the derivative of the function, $f(x_k)$ are the function values at those points, $A_{ik}^{(m)}$ are the weighting coefficients for the mth-order derivative of the function with respect to x, and N is the number of spatial grid points. Two points worth mentioning in the formulation of DQM are as follows: (1) how the weighting coefficients are determined; (2) how the grid points are selected.

The first step towards calculating the weighting coefficients is the selection of basis functions for approximating the unknown function $f(x)$ using the concept of linear vector space analysis and functional approximation. It is beyond the scope of this chapter to touch upon these fascinating fields of mathematics. The basics required for such an analysis are provided in the textbooks of DQM [32, 33]. There are various kinds of DQMs depending on the type of basis function used to calculate weighting coefficients. Generally used polynomial functions to compute the weighting coefficients are Legendre polynomials, Lagrange interpolation polynomials, Hermite polynomials, radial basis functions, and spline functions [41–45]. However, the most frequently used methods are based on Lagrange interpolation polynomials. In this work, weighting coefficients are calculated using Lagrange interpolation polynomials as the basis function. For a given set of N distinct points in the interval $[a, b]$, $a = x_1 < x_2 < x_3 \ldots < x_k \ldots < x_N = b$ and the corresponding value of $f(x_k)$, Lagrange's interpolation polynomials is a set of following $(N-1)$th degree polynomials:

$$L_k(x) = \frac{M(x)}{(x - x_k)M'(x_k)}, \quad k = 1, 2, 3, \ldots, N$$

where

$$M(x) = \prod_{j=1}^{N} (x - x_j), \quad M'(x_k) = \prod_{\substack{j=1 \\ j \neq k}}^{N} (x_k - x_j).$$

These polynomials constitute an N-dimensional vector space and are used as test functions to determine the weighting coefficients.

Using the polynomials, the interpolating function can be written as

$$f(x) \approx P_N(x) = \sum_{k=1}^{N} L_k(x) f(x_k).$$

Therefore, the first-order derivative of the function can be written as

$$f_x^1(x_i) = \left.\frac{\partial f(x)}{\partial x}\right|_{x=x_i} = \sum_{k=1}^{N} \left.\frac{\partial L_k(x)}{\partial x}\right|_{x=x_i} f(x_k), \quad i = 1, 2, \ldots, N. \tag{3}$$

Comparing Eqs. (2) and (3), we can write the weighting coefficients as

$$A_{ik}^{(1)} = \left.\frac{\partial L_k(x)}{\partial x}\right|_{x=x_i} = \frac{\partial}{\partial x}\left[\frac{M(x)}{(x - x_k)M'(x_k)}\right]_{x=x_i}.$$

Differentiating the Lagrange interpolation polynomial, we get

$$A_{ik}^{(1)} = \begin{cases} \dfrac{M'(x_i)}{(x_i - x_k)M'(x_k)} & \text{for } i \neq k = 1, 2, 3, \ldots, N \\ \displaystyle\sum_{\substack{k=1 \\ i\neq k}}^{N} A_{ik}^{(1)}. & \end{cases} \tag{4}$$

The weighting coefficients of the second-order derivative can be computed using

$$\left[A_{ik}^{(2)}\right] = \left[A_{ik}^{(1)}\right]\left[A_{ik}^{(1)}\right]. \tag{5}$$

Weighting coefficients of higher-order derivatives can be computed using the generalized formula,

$$A_{ik}^{(m)} = \begin{cases} m\left(A_{ii}^{(m-1)} A_{ik}^{(1)} - \dfrac{A_{ik}^{(m-1)}}{(x_i - x_k)}\right), & \text{for } i \neq k,\ i, k = 1, 2, \ldots, N \\ 2 \leq m \leq N-1 & \\ -\displaystyle\sum_{k=1.i\neq k}^{N} A_{ik}^{(m)}, & \text{for } i = k. \end{cases} \tag{6}$$

The proper selection of grid points provides the accuracy as well as the stability for any numerical method. Generally, grid points are based on the zeros of suitable orthogonal polynomials. Most widely used non-uniform grid points are Chebyshev–Gauss–Lobatto (CGL) grid points, and the formulation of the grid points as per CGL is given by

$$x_i = x_L + 0.5 \times \left(1 - \cos\frac{(i-1)\pi}{N-1}\right) \times L, \quad \text{for } i = 1, 2, \ldots, N, \tag{7}$$

where $L =$ length of the domain $= (x_R - x_L)$ for $x_L \leq x \leq x_R$.

3.2. *DQM-based Numerical Structure of the KdV Equation*

In this section, the KdV equation is discretized using the DQM formulation as given in Section 2. Replacing the first and third-order spatial derivative terms as present in the KdV equation with the corresponding algebraic terms of DQM, we get

$$\frac{\partial U(x,t)}{\partial t} + \varepsilon U(x,t)\frac{\partial U(x,t)}{\partial x} + \mu\frac{\partial^3 U(x)}{\partial x^3} = 0,$$

$$\frac{\partial U(x_i,t)}{\partial t} + \varepsilon U(x_i,t)\sum_{k=1}^{N} A_{ik}^{(1)} U(x_k,t) + \mu \sum_{k=1}^{N} A_{ik}^{(3)} U(x_k,t) = 0,$$
$$i = 1, 2, \ldots, N. \tag{8}$$

The time derivative term can be discretized using the finite difference scheme as

$$\frac{\partial U(x_i,t)}{\partial t} = \frac{U(x_i, t+\Delta t) - U(x_i,t)}{\Delta t},$$

where Δt is the discrete time step size. Therefore, Eq. (8) can be written as

$$\frac{U(x_i, t+\Delta t) - U(x_i,t)}{\Delta t} + \varepsilon U(x_i,t)\sum_{k=1}^{N} A_{ik}^{(1)} U(x_k,t)$$
$$+ \mu \sum_{k=1}^{N} A_{ik}^{(3)} U(x_k,t) = 0, \quad i = 1, 2, \ldots, N. \tag{9}$$

On rearranging Eq. (9), we get

$$U(x_i, t+\Delta t) = U(x_i, t) + \varepsilon U(x_i, t)\Delta t \sum_{k=1}^{N} A_{ik}^{(1)} U(x_k, t)$$
$$+ \mu\Delta t \sum_{k=1}^{N} A_{ik}^{(3)} U(x_k, t), \quad i = 1, 2, \ldots, N. \quad (10)$$

In matrix form, Eq. (10) can be written as

$$U_{t+1} = U_t + \varepsilon\Delta t U_t^T A^{(1)} U_t + \mu\Delta t A^{(3)} U_t,$$

where

$$U_{t+1} = \begin{pmatrix} U(x_1, t+\Delta t) \\ U(x_2, t+\Delta t) \\ \vdots \\ U(x_N, t+\Delta t) \end{pmatrix}, \quad U_t = \begin{pmatrix} U(x_1, t) \\ U(x_2, t) \\ \vdots \\ U(x_N, t) \end{pmatrix}.$$

U_t^T is the transpose of the column vector U_t, and $A^{(1)}$ and $A^{(3)}$ are $N \times N$ matrices defined respectively as

$$A^{(1)} = \begin{pmatrix} A_{11}^{(1)} & A_{12}^{(1)} & \cdots & A_{1N}^{(1)} \\ A_{21}^{(1)} & A_{22}^{(1)} & \cdots & A_{2N}^{(1)} \\ \vdots & \vdots & \vdots & \vdots \\ A_{N1}^{(1)} & A_{N1}^{(1)} & \cdots & A_{NN}^{(1)} \end{pmatrix} \quad \text{and}$$

$$A^{(3)} = \begin{pmatrix} A_{11}^{(3)} & A_{12}^{(3)} & \cdots & A_{1N}^{(3)} \\ A_{21}^{(3)} & A_{22}^{(3)} & \cdots & A_{2N}^{(3)} \\ \vdots & \vdots & \vdots & \vdots \\ A_{N1}^{(3)} & A_{N1}^{(3)} & \cdots & A_{NN}^{(3)} \end{pmatrix}.$$

4. Numerical Examples

In this section, the DQM-based numerical scheme for the KdV equation is verified by solving four standard benchmark problems, including single soliton, interaction of two solitons, interaction of three solitons, and single soliton with imprecise model parameter. The numerical solutions of these problems are compared with closed-form solutions, and the accuracy of the numerical scheme is measured by computing discrete root mean square error norm L_2 and maximum error norm L_∞ defined as

$$L_2 = \|U^{exact} - U^{dqm}\|_2 = \left[\frac{1}{N}\sum_{j=1}^{N}|U_j^{exact} - U_j^{dqm}|^2\right]^{\frac{1}{2}}$$

and

$$L_\infty = \|U^{exact} - U^{dqm}\|_\infty = \frac{max}{j}|U_j^{exact} - U_j^{dqm}|, \quad j = 1, 2, 3, \ldots, N,$$

respectively, where U^{exact} and U^{dqm} are the closed-form solution and DQM-based numerical solution, respectively.

4.1. *Single Soliton*

For the single soliton solution of the KdV equation, the following initial and boundary conditions are used:

$$U(x, t = 0) = 3c\,\mathrm{sech}^2(Ax + D),$$

$$U(x = x_L, t) = U(x = x_R, t) = 0,$$

where c is a constant associated with the velocity of the soliton as εc, A is a constant defined as

$$A = \frac{1}{2}\left(\frac{\varepsilon c}{\mu}\right)^{\frac{1}{2}},$$

D is also a constant, and x_L and x_R are the left and right boundary locations. Using the above initial and boundary conditions, the closed-form solution of the KdV equation is written as

$$U(x, t) = 3c\,\mathrm{sech}^2(Ax - Bt + D),$$

where B is a constant defined as

$$B = \varepsilon c A = \frac{1}{2}\varepsilon c\left(\frac{\varepsilon c}{\mu}\right)^{\frac{1}{2}}.$$

The above solution represents a single soliton moving to the positive x direction with velocity εc and amplitude $3c$. The single soliton motion is simulated by solving the DQM-based algebraic form of the KdV equation for the parameters $\varepsilon = 1$, $\mu = 4.84 \times 10^{-4}$, $c = 0.3$, $D = -6$, $x_1 = 0$, $x_2 = 4$. A total of 101 grid points is taken for this study. The time step is set at 0.005. Graphical representation of the position of the single soliton at times $t = 1, 2, 3, 4, 6$, and 8 as obtained from the closed-form and numerical solutions are shown in Figure 1. The error norms L_2 and L_∞ at times $t = 1, 4$, and 6 are shown in Table 1.

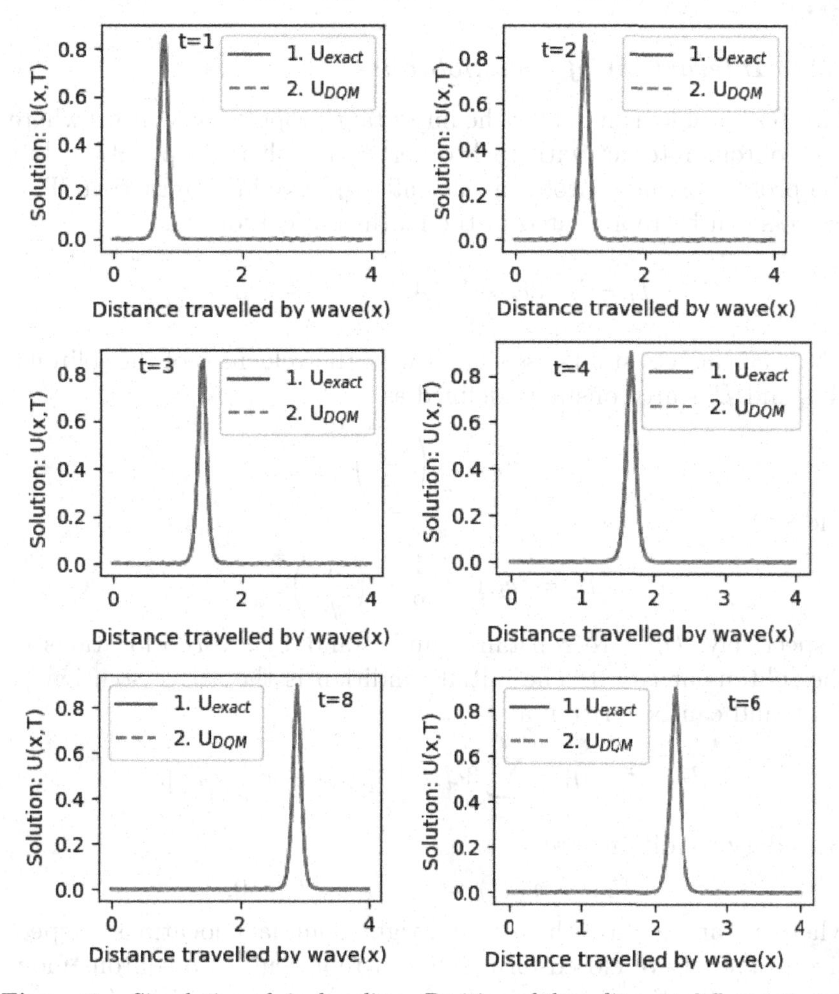

Figure 1. Simulation of single soliton: Position of the soliton at different times.

Table 1. L_2 and L_∞ error norms of the solution for single soliton.

Time	L_2	L_∞
1	0.002252	0.00626
2	0.002863	0.005841
3	0.002348	0.006173
4	0.002097	0.005509
6	0.002944	0.007723
8	0.002674	0.008688

4.2. *Interaction of Two Solitons*

This test problem simulates the important property of solitons where two solitons interact with each other during their propagation, but the profiles of the solitons are not affected by this interaction. This process can be represented in the mathematical form as

$$U(x,t) = \sum_{i=1}^{2} 3c_i \operatorname{sech}^2[A_i(x-x_i) - B_i t + D_i],$$

where c_i's are constants associated with the velocities of the solitons, A_i's and B_i's are constants defined as

$$A_i = \frac{1}{2}\left(\frac{\varepsilon c_i}{\mu}\right)^{\frac{1}{2}}$$

and

$$B_i = \varepsilon c_i A_i = \frac{1}{2}\varepsilon c_i \left(\frac{\varepsilon c_i}{\mu}\right)^{\frac{1}{2}},$$

respectively, D_i's are constants, and $x_1$and x_2 are the locations of the solitons at $t = 0$. The initial condition is the above solution at $t = 0$ and can be written as

$$U(x,t=0) = \sum_{i=1}^{2} 3c_i \operatorname{sech}^2[A_i(x-x_i) + D_i].$$

Boundary conditions are

$$U(x = x_L, t) = U(x = x_R, t) = 0,$$

where x_L and x_R are the left and right boundary locations, respectively. The above closed-form solution represents two solitons moving in the positive x direction with two different velocities $(\varepsilon c_1, \varepsilon c_2)$

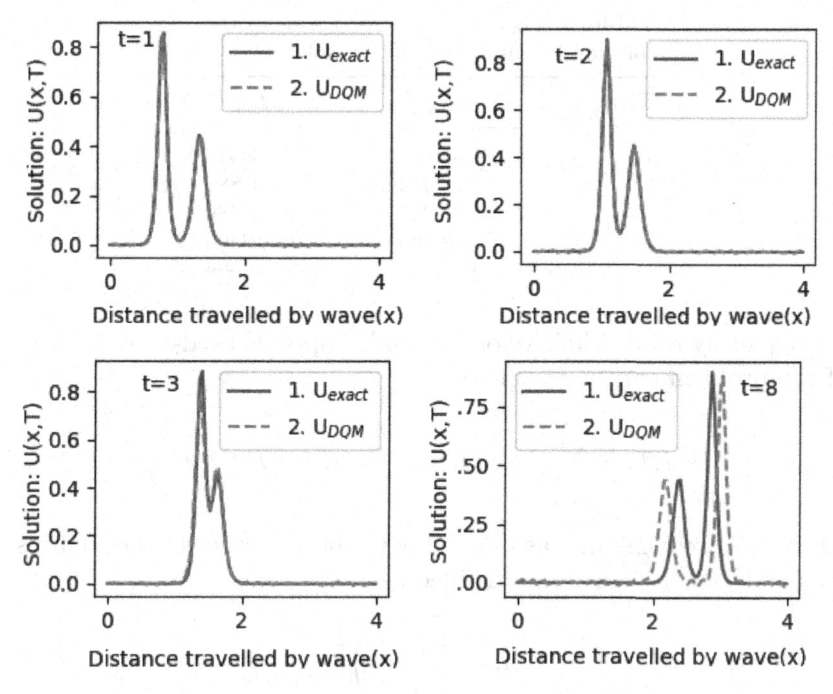

Figure 2. Simulation of the interaction of two solitons.

and amplitudes $(3c_1, 3c_2)$. The larger soliton is initially placed at $x = x_1 = 0$ and the smaller one at $x = x_2 = 0.5$. Since the larger soliton with amplitude $3c_1 = 3 \times 0.3 = 0.9$ has the larger velocity than the smaller one having amplitude $3c_2 = 3 \times 0.15 = 0.45$, it overtakes the smaller one at a certain time, with the velocities and shapes of both the solitons unaffected. The simulation is run with the parameters $\varepsilon = 1$, $\mu = 4.84 \times 10^{-4}$, $D = -6$, $x_L = 0$, $x_R = 4$. A total of 101 grid points is taken for this study. The time step is set at 0.005. Graphical representation of the two solitons obtained from the closed-form and numerical solutions at times $t = 1, 2, 3,$ and 8 are shown in Figure 2. The error norms L_2 and L_∞ at times $t = 1, 2, 3,$ and 8 are shown in Table 2.

4.3. *Interaction of Three Solitons*

In this test problem, three solitons moving in the positive x direction interact with each other due to their different velocities and

Table 2. L_2 and L_∞ error norms of the solution for interaction of two solitons.

Time	L_2	L_∞
1	0.002269	0.006318
2	0.003412	0.008885
3	0.022347	0.150480
8	0.199493	0.803403

subsequently retain their velocities and shapes. Closed-form solution of the problem can be written as

$$U(x,t) = \sum_{i=1}^{3} 3c_i \operatorname{sech}^2[A_i(x - x_i) - B_i t + D_i],$$

where c_i's are constants associated with the velocities of the solitons, A_i's and B_i's are constants defined as

$$A_i = \frac{1}{2}\left(\frac{\varepsilon c_i}{\mu}\right)^{\frac{1}{2}}$$

and

$$B_i = \varepsilon c_i A_i = \frac{1}{2}\varepsilon c_i \left(\frac{\varepsilon c_i}{\mu}\right)^{\frac{1}{2}},$$

respectively, D_i's are constants, and x_1, x_2, and x_3 are the locations of the solitons at $t = 0$. The initial condition is the above solution at $t = 0$ and can be written as

$$U(x, t = 0) = \sum_{i=1}^{3} 3c_i \operatorname{sech}^2[A_i(x - x_i) + D_i].$$

The boundary conditions are

$$U(x = x_L, t) = U(x = x_R, t) = 0,$$

where x_L and x_R are the left and right boundary locations, respectively. The above closed-form solution represents three solitons moving in the positive x direction with different velocities $(\varepsilon c_1, \varepsilon c_2, \varepsilon c_3)$ and amplitudes $(3c_1, 3c_2, 3c_3)$. The largest soliton is initially placed at

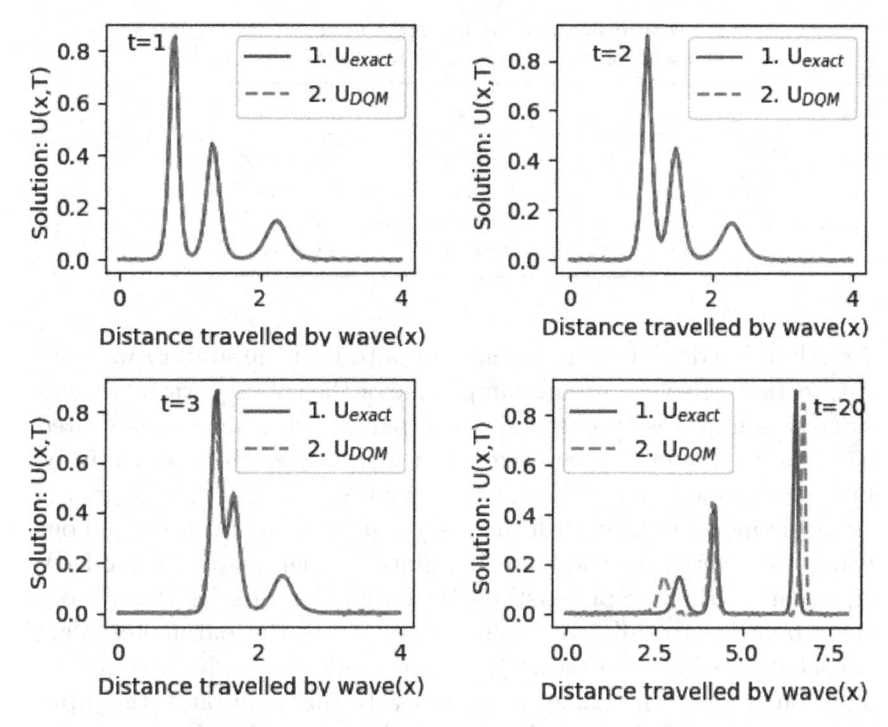

Figure 3. Simulation of interaction of three solitons.

$x = x_1 = 0$, and the other two solitons are placed at $x = x_2 = 0.5$ and $x = x_3 = 1.0$. The amplitudes of the solitons are $3c_1 = 3 \times 0.3 = 0.9$, $3c_2 = 3 \times 0.15 = 0.45$, and $3c_3 = 3 \times 0.05 = 0.15$. During propagation, they interact with each other without affecting their velocities and shapes. The simulation is run with the parameters $\varepsilon = 1$, $\mu = 4.84 \times 10^{-4}$, $D = -6$, $x_L = 0$, $x_R = 6$ and 8. Grid size for this study is fixed at 0.04, and the time step is set at 0.005. Graphical results of the three solitons before and after the interaction obtained from the closed-form and numerical solutions at times $t = 1, 2, 3$, and 20 are shown in Figure 3. The error norms L_2 and L_∞ at times $t = 1, 2, 3$, and 20 are shown in Table 3.

4.4. *Single Soliton with Imprecise Parameters*

In Section 4.1, the closed-form solution of the KdV equation for a single soliton was solved for a fixed value of input parameter, c, which is

Table 3. L_2 and L_∞ error norms of the solution for interaction of three solitons.

Time	L_2	L_∞
1	0.00182	0.005825
2	0.002636	0.011337
3	0.018321	0.150671
20	0.146061	0.880676

directly related to the velocity and amplitude of the solitary wave. In this section, the velocity and amplitude of the solitary wave are considered as imprecise parameters. The parameters ε and μ associated with the KdV equation are fixed at 1 and 4.84×10^{-4}, respectively, but the parameter c is treated as a fuzzy variable. Uncertainty analysis of the model output with the fuzzy input parameter is carried out using fuzzy vertex method [46–48]. Since the parameters of the KdV equation (ε, μ) are kept constant, the only different initial conditions corresponding to different alpha-cut values of the parameter c are considered as input to the DQM-based model of the KdV equation. The complete methodology is started with fuzzification of the input parameters, c, of the closed-form solution of the KdV equation for the single soliton. Fuzziness of the parameters is expressed in terms of a triangular fuzzy number, with lower bound, most likely, and upper bound values being 0.2, 0.25, and 0.3, respectively. The numerical values of the other parameters associated with the closed-form solution for single soliton are the same as those used in Section 4.1. The minimum and maximum of the solution at any spatial and temporal coordinate for all the values of alpha ranging from 0 to 1 with an increment of 0.1 are constructed. Both closed-form and DQM-based solutions, $U(xt)$, for alpha-cut values of 0.0, 0.1, 0.5, and 0.9 at time 5 are shown in Figure 4.

5. Conclusions

Polynomial DQM has been applied to obtain a numerical solution of the KdV equation, which mathematically represents shallow water waves. For single soliton, the root mean square error norm L_2 and maximum error norm L_∞ are very small, and the graphical results

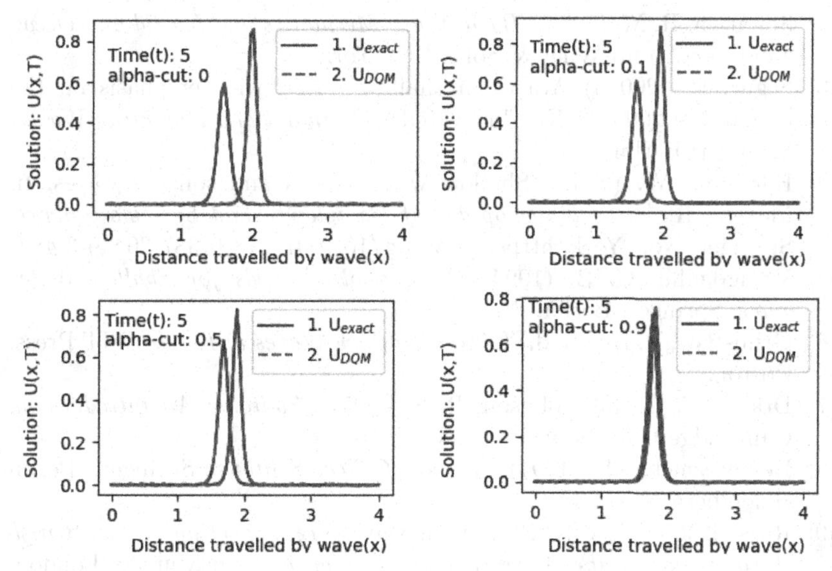

Figure 4. Solutions of the KdV equation with imprecise model parameters at various alpha-cuts.

also show that a very accurate solution can be obtained using this technique. For the problems of interaction between two and three solitons, the numerical results are very accurate before their interaction but are slightly shifted after the interaction. Nevertheless, the solutions for the cases are highly stable with time. The imprecise parameter associated with the velocity and amplitude of the soliton is treated as a triangular fuzzy variable, and the corresponding membership function of the solution of the KdV equation is estimated and compared with the closed-form solution and highly accurate results are obtained. The alpha-cut value of a fuzzy number being an interval, the lower and upper bounds are independently adopted for obtaining the numerical solution of the KdV equation.

References

[1] Dean, R. G. and Dalrymple, R. A. (1991). *Water Wave Mechanics for Engineers and Scientists*, Advanced Series on Ocean Engineering, Vol. 2, World Scientific: Singapore.

[2] Sorensen, R. M. (1993). *Basic Wave Mechanics for Coastal and Ocean Engineers*, John Wiley & Sons: New York.
[3] Segur, H. (2007a). Waves in shallow water, with emphasis on the tsunami of 2004. In Kundu, A. (ed.), *Tsunami and Nonlinear Waves*, Springer: Berlin.
[4] Hereman, W. (2009). Shallow water waves and solitary waves. In Meyers, R. (ed.), *Encyclopedia of Complexity and Systems Science*, Springer: New York. https://doi.org/10.1007/978-0-387-30440-3_480.
[5] Vreugdenhil, C. B. (1994). *Numerical Methods for Shallow-Water Flow*. Springer: Berlin.
[6] Grimshaw, R. H. J. (ed.) (2007). *Solitary Waves in Fluids*. WIT Press: Boston.
[7] Drazin, P. G. and Johnson, R. S. (1996). *Solitons: An Introduction*. Cambridge University Press: New York.
[8] Remoissenet, M. (1994). *Waves Called Solitons*. Springer: Berlin Heidelberg New York.
[9] Russell, J. S. (1844). Report on waves, *14th meeting of the British Association for the Advancement of Science*. John Murray: London, pp. 311–390.
[10] Scott, A. C. *et al.* (1973) The soliton: A new concept in applied science. *Proceedings of the IEEE*, 61, pp. 1443–1483.
[11] Miura, R. M. (1976). The Korteweg–deVries equation: A survey of results. *SIAM Review*, 18(3), 412–459.
[12] Zabusky, N. J. and Kruskal, M. D. (1965). Interaction of 'solitons' in a collisionless plasma and the recurrence of initial states. *Physics Review Letters*, 15, 240–243.
[13] Das, G. C. and Sarma, J. (1998). A new mathematical approach for finding the solitary waves in dusty plasma. *Physics of Plasmas*, 5(11), 3918–3923.
[14] Osborne, A. R. (1995). The inverse scattering transform: Tools for the nonlinear Fourier analysis and filtering of ocean surface waves. *Chaos, Solitons & Fractals*, 5(12), 2623–2637.
[15] Ludu, A. and Jerry, P. D. (1998). Nonlinear modes of liquid drops as solitary waves. *Physical Review Letters*, 80(10), 2125.
[16] Turitsyn, S. K., Aceves, A. B., Jones, C. K. R. T., and Zharnitsky, V. (1998). Average dynamics of the optical soliton in communication lines with dispersion management: Analytical results. *Physical Review E*, 58(1), R48.
[17] Coffey, M. W. (1996). Nonlinear dynamics of vortices in ultraclean type-II superconductors: Integrable wave equations in cylindrical geometry. *Physical Review B*, 54(2), 1279.
[18] Benjamin, T. B., Bona, J. L., and Mahony, J. J. (1972). Model equations for long waves in nonlinear dispersive systems. *Philosophical*

Transactions of the Royal Society of London. Series A, Mathematical and Physical Sciences, 272(1220), 47–78.

[19] Korteweg, D. G. and de Vries, G. (1895). On the change of form of long waves advancing in a rectangular canal, and on a new type of long stationary wave, *The London, Edinburgh, and Dublin Philosophical Magazine and Journal of Science*, 39(1895), 422–443.

[20] Clamond, D. (2003). Cnoidal-type surface waves in deep water. *Journal of Fluid Mechanics*, 489, 101–120.

[21] Soliman, A. A. and Abdou, M. A. (2007). Exact travelling wave solutions of nonlinear partial differential equations. *Chaos, Soliton & Fractals*, 32(2), 808–815.

[22] Gardner, C. S., Green, J. M., Kruskal, M. D., and Miura, R. M. (1967). Method for Solving the Korteweg–de Vries Equation. *Physical Review Letters*, 19, 1095–1097.

[23] Hirota, R. (1971). Exact solution of the Korteweg–de Vries equation for multiple collisions of solitons. *Physical Review Letters*, 27(18), 1192.

[24] Hassan, I. H. A. H. (2008). Comparison differential transformation technique with Adomian decomposition method for linear and nonlinear initial value problems. *Chaos, Solitons & Fractals*, 36(1), 53–65.

[25] Helal, M. A. and Mehanna, M. S. (2006). A comparison between two different methods for solving KdV–Burgers equation. *Chaos, Solitons & Fractals*, 28(2), 320–326.

[26] Oruç, Ö., Bulut, F., and Esen, A. (2016). Numerical solution of the KdV equation by Haar wavelet method. *Pramana*, 87(6), 1–11.

[27] Khattak, A. J. and Tirmizi, I. A. (2008). A meshfree method for numerical solution of KdV equation. *Engineering Analysis with Boundary Elements*, 32(10), 849–855.

[28] Gardner, G. A., Ali, A. H. A., and Gardner, L. R. T. (1989). A finite element solution for the Korteweg–de Vries equation using cubic B-spline shape functions. In *ISNME-89*, Springer: Berlin, 2, pp. 565–570.

[29] Dağ, İ. and Yılmaz D. (2008). Numerical solutions of KdV equation using radial basis functions. *Applied Mathematical Modelling*, 32(4), 535–546.

[30] Soliman, A. A. (2004). Collocation solution of the Korteweg–de Vries equation using septic splines. *International Journal of Computer Mathematics*, 81(3), 325–331.

[31] Bert C. W. and Malik, M. (1996). Differential quadrature method in computational mechanics: A review. *Applied Mechanics Reviews*, 49(1), 1–28.

[32] Shu, C. (2000). *Differential Quadrature and Its Application in Engineering*. Springer: Berlin.

[33] Zong, Z. and Zhang, Y. (2009). *Advanced Differential Quadrature Methods*, CRC Press: New York.

[34] Wu, T. Y. and Liu, G. R. (1999). A differential quadrature as a numerical method to solve differential equations. *Computational Mechanics*, 24(3), 197–205.

[35] Meral, G. (2013). Differential quadrature solution of heat and mass-transfer equations. *Applied Mathematical Modelling*, 37(6), 4350–4359.

[36] de Falco, M., Gaeta, M., Loia, V., Rarità, L., and Tomasiello, S. (2016). Differential quadrature-based numerical solutions of a fluid dynamic model for supply chains. *Communications in Mathematical Sciences*, 14(5), 1467–1476.

[37] Kaya, B. and Arisoy, Y. (2011). Differential quadrature solution for one-dimensional aquifer flow. *Mathematical and Computational Applications*, 16(2), 524—534.

[38] Kaya, B. (2010). Solution of the advection–diffusion equation using the differential quadrature method. *KSCE Journal of Civil Engineering*, 14(1), 69–75.

[39] Zhu, X. G., Nie, Y. F., and Zhang, W. W. (2017). An efficient differential quadrature method for fractional advection–diffusion equation. *Nonlinear Dynamics*, 90(3), 1807–1827.

[40] Datta, D. and Pal, T. K. (2017). Development of fuzzy differential quadrature numerical method and its application for uncertainty quantification of solute transport model. *Life Cycle Reliability and Safety Engineering*, 6(4), 249–256.

[41] Bellman, R. E. and Casti, J. (1971). Differential quadrature and long-term integration. *Journal of Mathematical Analysis and Applications*, 34, 235–238.

[42] Bellman, R. E., Kashef, B. G., and Casti, J. (1972). Differential quadrature: A technique for the rapid solution of nonlinear partial differential equations. *Journal of Computational Physics*, 10, 40–52.

[43] Cheng, J., Wang, B., and Du, S. Y. (2005). A theoretical analysis of piezoelectric/composite laminate with larger-amplitude deflection effect, Part II: Hermite differential quadrature method and application. *International Journal of Solids and Structures*, 42(24–25) 6181–6201.

[44] Shu, C. and Wu, Y. L. (2007). Integrated radial basis functions-based differential quadrature method and its performance. *International Journal for Numerical Methods in Fluids*, 53(6) 969–984.

[45] Striz, A. G., Wang, X., and Bert, C. W. (1995). Harmonic differential quadrature method and applications to analysis of structural components. *Acta Mechanica*, 111(1), 85–94.

[46] Klir, G. J. and Yuan, B. (1995). *Fuzzy sets and Fuzzy Logic: Theory and Applications*, Prentice Hall, PTR: Upper Saddle River, NJ.
[47] Dubois, D. and Prade, H. (1987). Fuzzy sets and statistical data, *European Journal of Operational Research*, 25, 345–356.
[48] Dong, W. and Shah, H. (1987), Vertex method for computing functions of fuzzy variables. *Fuzzy Sets and Systems*, 24(1), 65–78.

https://doi.org/10.1142/9789811245367_0005

Chapter 5

Seismic Waves and Their Effect on Structures

Sagarika Mukhopadhyay* and Akash Kharita

Department of Earth Sciences, IIT Roorkee, Roorkee 247667, Uttarakhand, India

**sagarika.mukhopadhyay@es.iitr.ac.in*

Abstract

Seismic waves are mechanical vibrations that propagate through the Earth. They cannot propagate through vacuum. There are various types of seismic waves (e.g., body and surface waves); each type can be categorized into two subtypes based on the nature of particle motion during wave propagation. The two body wave types are P wave and S wave. The P wave is also known as primary wave or longitudinal wave. Particle motion or the oscillation of the medium during the propagation of this wave is in the direction of wave propagation; hence, it is similar to a sound wave. During wave propagation, the body experiences compression and dilatation, i.e., change in volume. This wave can propagate through both solid and fluid media and is the fastest of all seismic waves. The S wave is also known as shear wave or secondary wave. The medium oscillates in a direction perpendicular to the direction of propagation of wave front. The body experiences shearing motion that leads to a change in its shape, but no change in volume. As fluids cannot sustain shear, S waves cannot move through them. Hence, an S wave propagates only through solid media. In a given medium, its velocity is lower than that of a P wave; hence, it always arrives after the P wave, and that is why it is also called as secondary wave.

Surface waves develop due to interference of post-critical reflected P and S waves. Their maximum amplitude of vibration decreases exponentially with depth in the Earth. Hence, their propagation effect is maximum near the surface. That is why they are called surface wave. They travel with velocity less than that of S wave and arrive later than body waves. The two types of surface waves are known as Rayleigh wave and Love wave. A Rayleigh wave is generated by the constructive interference of a P wave and the vertical component of an S wave (also called SV wave). In a record of seismic wave called seismogram, normally this is the maximum amplitude at arrival. A Love wave is generated by the constructive interference of upgoing and downgoing components of an SH wave (where particle motion of the S wave is in the horizontal direction). Love waves travel faster than Rayleigh waves. There is another type of wave similar to the surface waves in nature that travels along an interface at deeper locations within the Earth called Stonley wave.

Apart from this, the Earth also experiences free oscillations — standing waves generated due to interference of long-period seismic waves. These waves cause the whole Earth to vibrate. There are two types of free oscillations: spheroidal and toroidal. Spheroidal oscillation is similar to the motion caused by a Rayleigh wave, and Toroidal motion is similar to the motion caused by a Love wave.

Keywords: Seismic wave, velocity, wave equation, Ray path, travel time, reflection and transmission coefficients

1. Introduction

All the above-mentioned types of seismic waves can be used to study the interior of the Earth. Based on several such studies, the Earth's interior can be represented by a few near-spherical concentric shells, *viz.*, crust, mantle, outer core, and inner core. Seismic wave velocity usually increases with depth. This is the first-order representation of the Earth's seismic wave velocity structure, where it is assumed that at the same depth, velocity everywhere is the same. This assumption is largely applicable because the rate of variation of seismic wave velocity in the depth direction is much larger than that in the lateral direction. However, with the advent of abundant records of seismic waves, the modern approach is to find 3D variation in seismic wave velocity, normally known as seismic tomography. Information gleaned from such studies is used to understand the manifestation of geologic, tectonic, and geodynamic processes that have shaped the Earth and reveal the nature of medium within the Earth.

In order to carry out such investigations, one should understand the basic mechanism of seismic wave propagation in the Earth. This requires a firm grasp of the wave equation as applicable to seismic waves, which are nothing but mechanical vibrations. To a large extent, the Earth behaves like an elastic body. Hence, in this chapter, the derivation of the seismic wave equation and its solution will be carried out for an elastic medium. Solution of wave equation for both homogeneous and heterogeneous media will be derived, and from them, travel time equations along ray paths that obey Snell's law will be obtained for simple velocity structures. For derivation of seismic wave equation for an elastic medium, elastic stress–strain relation for a 3D medium is necessary. Hence, in this chapter, a concise description of such a relation is given at the beginning. As the Earth has regions where velocity changes slowly and other regions where there is sudden change in velocity, seismic waves can travel from source to recording stations along multiple paths. In this chapter, travel time equations for both types of media will be derived. The wave amplitudes for waves traveling along such paths are also affected by such variations in velocity. At the end of the chapter, how wave amplitude is affected by velocity variation will be discussed.

2. Theory of Elasticity in 3D

When seismic waves propagate through a part of the Earth, it causes strain that varies with time. Away from the source, this strain is normally very small so that it may be considered to be linearly dependent on the causative stress (Eq. (1)). This gives us Hooke's law in 3D [1]:

$$\tau_{ij} = C_{ijkl} e_{kl}, \tag{1}$$

where value of each suffix i, j, k, l varies between $1, 2, 3$, and represent the directions x_1, x_2, and x_3, respectively (Figure 1), τ_{ij} is the stress tensor, e_{kl} is the strain tensor, and C_{ijkl} is the elastic constant. The first subscript in stress tensor represents the direction perpendicular to the surface on which the stress is acting, and the second one represents the direction in which the stress is acting (Figure 1). When $i = j$, the stress acts normal to the corresponding surface and hence it is called normal stress, and when $i \neq j$, it acts parallel to the surface and is called shear stress (Figure 1). In the same fashion, strain tensors are classified as normal and shear strains. From Figure 1, it is observed that there can be nine stress components.

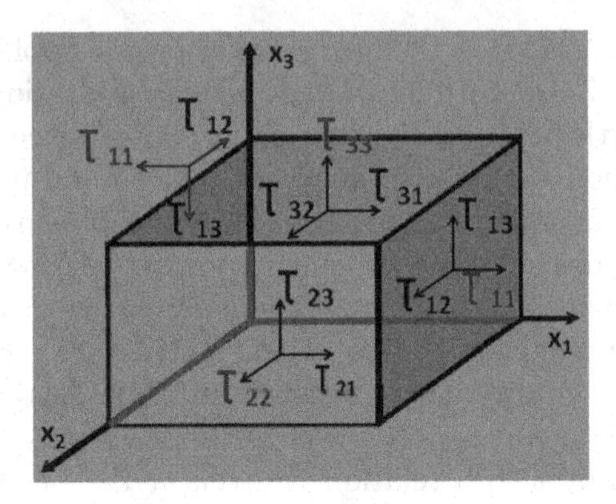

Figure 1. Schematic representation of stress tensors.

Similarly, we can draw a figure like Figure 1 and show that there can be nine strain components. This means there can be 81 independent elastic constants C_{ijkl} for a given body.

Under static condition, i.e., when the body is not spinning, $\tau_{ij} = \tau_{ji}$ and $e_{kl} = e_{lk}$. As a consequence, there can be only six independent stress components as well as six independent strain components. Consequently, a body having maximum anisotropy should be represented by 36 independent elastic constants. However, strain energy W satisfies the following condition, leading to the situation that maximum number of elastic constants required to describe a completely anisotropic body becomes 21:

$$W = {}^1\!/_2 \tau_{ij} e_{kl} = {}^1\!/_2 \tau_{kl} e_{ij} \tag{2}$$

If the body has one plane of symmetry, then the number of independent elastic constants reduces to 15. If it has three planes of symmetry, it further reduces to nine. Such a body is called an orthotropic body. An example of such a body is wood. For transversely isotropic material, i.e., when the body is symmetric around an axis, the number of elastic constants required is five. There are numerous examples of such bodies found on Earth, such as finely layered rocks, rocks made of anisotropic crystals with preferred orientation, and rocks having sets of joints/fractures in one direction. When a body is completely isotropic, its elastic properties can be represented by two independent elastic constants λ and μ, called Lame's parameters. μ is also called shear modulus. Three other elastic parameters, *viz.*, Young's modulus E, Poisson's ratio σ, and bulk modulus k are related to these parameters through the following equations:

$$E = \mu(3\lambda + 2\mu)/(\lambda + \mu), \tag{3}$$

$$\sigma = \lambda/2(\lambda + \mu), \tag{4}$$

$$k = (3\lambda + 2\mu)/3. \tag{5}$$

For an isotropic medium, the Hooke's law can be represented as

$$\begin{aligned} \tau_{ij} &= \lambda\theta\delta_{ij} + 2\mu e_i \\ &= \lambda\theta + 2\mu e_{ij} \quad (\text{when } i = j), \\ &= 2\mu e_{ij} \quad (\text{when } i \neq j), \end{aligned} \tag{6}$$

where volume strain $\theta = e_{11} + e_{22} + e_{33}$ and δ_{ij} is the Kroneker delta. If u_i is the displacement in x_i direction, where i varies between 1 and 3 at an instant of time t when seismic wave propagates through a medium, then corresponding strain e_{ij} is given by (subscripts i and j represent directions x_1, x_2 and x_3)

$$e_{ij} = \tfrac{1}{2}(\partial u_i/\partial x_j + \partial u_j/\partial x_i) \tag{7a}$$

$$\text{and } \theta = \partial u_1/\partial x_1 + \partial u_2/\partial x_2 + \partial u_3/\partial x_3 \tag{7b}$$

In this chapter, the discussion will be confined to the wave equation for an elastic and isotropic medium. In the next section, using basic principles of Physics, the wave equation for seismic waves will be derived.

3. Derivation of Seismic Wave Equation Using Basic Principles of Physics

3.1. *Equation of Equilibrium*

Let F_i and T_i be the body and surface forces per unit volume and surface, respectively, acting in the ith direction on a body. Here, the body force represents the force acting within the volume of a material, and the surface force represents the force acting on the surface of the body. If the body is not moving under the action of these forces, then

$$\int_v F_i\,dv + \int_s \tau\gamma_i\,ds = 0. \tag{8}$$

Here, the net force is zero, and γ represents direction cosine. Here, the first term on the left-hand side represents volume integral and the second one represents surface integral. The surface force T_i can be represented as (here, the convention is when a subscript is repeated in an equation, it means summation is used)

$$T_i = \tau_{ji}\gamma_j = \tau_{1i}\gamma_1 + \tau_{2i}\gamma_2 + \tau_{3i}\gamma_3 \tag{9}$$

Using divergence theorem, Eq. (8) can be written as

$$\int_v (F_i + \tau_{ji,j})dv = 0, \tag{10}$$

where $\tau_{ji,j} = \partial\tau_{ji}/\partial x_j$. Hence, for the static case, i.e., when the body is not moving, we get

$$\frac{\partial\tau_{11}}{\partial x_1} + \frac{\partial\tau_{21}}{\partial x_2} + \frac{\partial\tau_{31}}{\partial x_3} = -F_1, \tag{11a}$$

$$\frac{\partial\tau_{12}}{\partial x_1} + \frac{\partial\tau_{22}}{\partial x_2} + \frac{\partial\tau_{32}}{\partial x_3} = -F_2, \tag{11b}$$

$$\frac{\partial\tau_{13}}{\partial x_1} + \frac{\partial\tau_{23}}{\partial x_2} + \frac{\partial\tau_{33}}{\partial x_3} = -F_3. \tag{11c}$$

For the dynamic case, i.e., when the body is moving, then according to Newton's second law of motion we get

$$F_i + \tau_{ji,j} = \rho\frac{\partial^2 u_i}{\partial t^2}. \tag{12a}$$

If $i = 1$, we get

$$F_1 + \tau_{j1,j} = \rho\frac{\partial^2 u_1}{\partial t^2}. \tag{12b}$$

Using Eqs. (6) and (7), the above equation can be written as follows:

$$\lambda\frac{\partial\theta}{\partial x_1} + 2\mu\frac{\partial e_{j1}}{\partial x_j} + F_1 = \rho\frac{\partial^2 u_1}{\partial t^2}, \tag{13a}$$

$$\text{or}\quad \lambda\frac{\partial\theta}{\partial x_1} + 2\mu\left(\frac{\partial e_{11}}{\partial x_1} + \frac{\partial e_{21}}{\partial x_2} + \frac{\partial e_{31}}{\partial x_3}\right) + F_1 = \rho\frac{\partial^2 u_1}{\partial t^2}, \tag{13b}$$

$$\text{or}\quad \lambda\frac{\partial\theta}{\partial x_1}+2\mu\left[\frac{\partial^2 u_1}{\partial x_1^2}+{}^1\!/\!_2\frac{\partial}{\partial x_2}\left(\frac{\partial u_2}{\partial x_1}+\frac{\partial u_1}{\partial x_2}\right)\right.$$

$$\left.+{}^1\!/\!_2\frac{\partial}{\partial x_3}\left(\frac{\partial u_3}{\partial x_1}+\frac{\partial u_1}{\partial x_3}\right)\right]+F_1=\rho\frac{\partial^2 u_1}{\partial t^2}, \tag{13c}$$

$$\text{or}\quad \lambda\frac{\partial\theta}{\partial x_1}+\mu\left[\frac{\partial}{\partial x_1}\left(\frac{\partial u_1}{\partial x_1}+\frac{\partial u_2}{\partial x_2}+\frac{\partial u_3}{\partial x_3}\right)+{}^1\!/\!_2\frac{\partial}{\partial x_3}\left(\frac{\partial u_3}{\partial x_1}+\frac{\partial u_1}{\partial x_3}\right)\right.$$

$$\left.+\frac{\partial^2 u_1}{\partial x_1^2}+\frac{\partial^2 u_1}{\partial x_2^2}+\frac{\partial^2 u_1}{\partial x_2^3}\right]+F_1=\rho\frac{\partial^2 u_1}{\partial t^2}, \tag{13d}$$

$$\text{or}\quad (\lambda+\mu)\frac{\partial\theta}{\partial x_1}+\mu\nabla^2 u_1+F_1=\rho\frac{\partial^2 u_1}{\partial t^2}$$

$$=\text{net unbalanced force in } x_1 \text{ direction.} \tag{13e}$$

By analogy, the net unbalanced force in x_2 and x_3 directions are

$$(\lambda+\mu)\frac{\partial\theta}{\partial x_2}+\mu\nabla^2 u_2+F_2=\rho\frac{\partial^2 u_2}{\partial t^2}, \tag{13f}$$

$$(\lambda+\mu)\frac{\partial\theta}{\partial x_3}+\mu\nabla^2 u_3+F_3=\rho\frac{\partial^2 u_3}{\partial t^2}. \tag{13g}$$

3.2. *Wave Equation in a Source-free Region*

For seismic waves, the body force is present only when an earthquake or explosion or implosion source is present. In a source-free region, $F_i = 0$. Putting this in Eqs. (13(e) and 13(f)), differentiating them with respect to x_1, x_2, and x_3, and adding them we get

$$(\lambda+\mu)\left(\frac{\partial^2\theta}{\partial x_1^2}+\frac{\partial^2\theta}{\partial x_2^2}+\frac{\partial^2\theta}{\partial x_3^2}\right)+\mu\nabla^2\left(\frac{\partial u_1}{\partial x_1}+\frac{\partial u_2}{\partial x_2}+\frac{\partial u_3}{\partial x_3}\right)$$

$$=\rho\frac{\partial^2}{\partial t^2}\left(\frac{\partial u_1}{\partial x_1}+\frac{\partial u_2}{\partial x_2}+\frac{\partial u_3}{\partial x_3}\right), \tag{14a}$$

$$\text{or}\quad (\lambda+2\mu)\nabla^2\theta=\rho\frac{\partial^2\theta}{\partial t^2}, \tag{14b}$$

$$\text{or}\quad \nabla^2\theta=\frac{\rho}{(\lambda+2\mu)}\frac{\partial^2\theta}{\partial t^2}=\frac{1}{\alpha^2}\frac{\partial^2\theta}{\partial t^2}, \tag{14c}$$

where α is the velocity with which the P wave travels in a medium. Equation (14(c)) shows that when P wave propagates through a medium, it leads to a change in volume/length.

Putting $F_3 = 0$ in Eq. (13(g)), taking its derivative with respect to x_2, and then, taking $F_2 = 0$ in Eq. (13(f)), taking its derivative with respect to x_3, and subtracting it from the first one we get

$$(\lambda+\mu)\frac{\partial^2\theta}{\partial x_2 \partial x_3} + \mu\nabla^2\frac{\partial u_3}{\partial x_2} - (\lambda+\mu)\frac{\partial^2\theta}{\partial x_2 \partial x_3} - \mu\nabla^2\frac{\partial u_2}{\partial x_3}$$

$$= \rho\frac{\partial^2}{\partial t^2}\left(\frac{\partial u_3}{\partial x_2} - \frac{\partial u_2}{\partial x_3}\right), \tag{15a}$$

$$\text{or}\quad \mu\nabla^2\left(\frac{\partial u_3}{\partial x_2} - \frac{\partial u_2}{\partial x_3}\right) = \rho\frac{\partial^2}{\partial t^2}\left(\frac{\partial u_3}{\partial x_2} - \frac{\partial u_2}{\partial x_3}\right), \tag{15b}$$

$$\text{or}\quad \mu\nabla^2\theta_{x1} = \rho\frac{\partial^2\theta_{x1}}{\partial t^2}, \tag{15c}$$

$$\text{or}\quad \nabla^2\theta_{x1} = \frac{\rho}{\mu}\frac{\partial^2\theta_{x1}}{\partial t^2}, \tag{15d}$$

$$\text{or}\quad \nabla^2\theta_{x1} = \frac{1}{\beta^2}\frac{\partial^2\theta_{x1}}{\partial t^2}, \tag{15e}$$

where θ_{x_1} represents rotation in x_2x_3 plane perpendicular to x_1 axis and β is the S wave velocity. This means that when the S wave propagates through a medium, it leads to rotation or shear of the medium.

3.3. *Wave Equation in Vector Form*

Wave equations for P and S waves, given in Eqs. (14) and (15), respectively, can also be represented in vector form, as shown in the following equation, with displacement $\bar{u}$ being a vector:

$$\rho\ddot{\bar{u}} = (\lambda+\mu)\nabla(\nabla\cdot\bar{u}) + \mu\nabla^2\bar{u}, \tag{16}$$

where $\ddot{\bar{u}}$ represents the second order differential of $\bar{u}$ with respect to time t. Now,

$$\nabla^2\bar{u} = \nabla(\nabla\cdot\bar{u}) - (\nabla X \nabla X \bar{u}). \tag{17}$$

Therefore, Eq. (16) can be written as

$$\rho\ddot{\bar{u}} = (\lambda + 2\mu)\nabla(\nabla \cdot \bar{u}) - \mu(\nabla X \nabla X \bar{u}). \tag{18}$$

According to Helmholtz's theorem, any vector $\bar{u}$ can be represented as

$$\bar{u} = \nabla\Phi + \nabla X \Psi, \tag{19}$$

where Φ is the scalar potential and Ψ is the vector potential. Hence, Φ is curl free and Ψ is divergence free. So, we can write $\nabla X \Phi = 0$ and $\nabla \cdot \Psi = 0$. Putting Eq. (19) in Eq. (18), we get

$$\begin{aligned}\rho\frac{\partial^2}{\partial t^2}[\nabla\Phi + \nabla X \Psi] &= (\lambda + 2\mu)\nabla[\nabla \cdot (\nabla\Phi + \nabla X \Psi)] \\ &\quad - \mu\nabla X \nabla X[\nabla\Phi + \nabla X \Psi]. \end{aligned} \tag{20}$$

From the definitions of Φ and Ψ, we get $\nabla X \nabla X \Psi = -\nabla^2\Psi$ and $\nabla(\nabla \cdot \Phi) = \nabla^2\Phi$. Putting this in Eq. (20), we get

$$\nabla\left[(\lambda + 2\mu)\nabla^2\Phi - \rho\frac{\partial^2\Phi}{\partial t^2}\right] + \nabla X\left[\mu\nabla^2\Psi - \rho\frac{\partial^2\Psi}{\partial t^2}\right] = 0. \tag{21a}$$

Equation (21(a)) shows that the sum of divergence of some parameters and curl of some parameters is equal to zero. In that case, individually, those parameters should be zero. Therefore,

$$\nabla^2\Phi = \frac{1}{\alpha^2}\frac{\partial^2\Phi}{\partial t^2}, \tag{21b}$$

and

$$\nabla^2\Psi = \frac{1}{\beta^2}\frac{\partial^2\Psi}{\partial t^2}. \tag{21c}$$

Comparing Eqs. (14(c)) and (21(b)), it can be said that the P wave motion can be represented by a scalar potential. Similarly, comparing Eqs. (14(d)) and (21(c)), it can be said that the S wave motion can be represented by a vector potential. In the next section, the derivation of the solutions of the wave equation for various types of media is discussed.

4. Solution for Wave Equation

4.1. *D' Alembert's Solution for Plane Wave*

Let there be a plane wave front propagating in x_1 direction. Here, $u_2 = u_3 = 0$, and the wave equation can be written as

$$\frac{\partial^2 u_1}{\partial x_1^2} = \frac{1}{V^2}\frac{\partial^2 u_1}{\partial t^2}, \tag{22}$$

where V represents seismic wave velocity. Let the medium have a constant velocity. For the P wave, $V = \alpha$, and for the S wave, $V = \beta$. The general solution or D'Alembert's solution of this equation can be written as

$$u_1(x_1, t) = f(x_1 - Vt) + g(x_1 + Vt). \tag{23}$$

Equation (22) is a partial differential equation that can be solved using separation of variable method, assuming the solution has a form that separates spatial and temporal dependences. Let the solution be of the form

$$u_1(x_1, t) = X(x_1)T(t). \tag{24}$$

Putting this in Eq. (22), we get

$$\frac{X(x_1)}{V^2}\frac{d^2T(t)}{dt^2} = T(t)\frac{d^2X(x_1)}{dx_1^2}, \tag{25a}$$

$$\text{or}\quad \frac{1}{T(t)}\frac{d^2T(t)}{dt^2} = \frac{V^2}{X(x_1)}\frac{d^2X(x_1)}{dx_1^2} = -\omega^2\,(constant). \tag{25b}$$

Therefore,

$$\frac{d^2X(x_1)}{dx_1^2} + \frac{\omega^2}{V^2}X(x_1) = 0 \tag{25c}$$

$$\frac{d^2T(t)}{dt^2} + \omega^2T(t) = 0. \tag{25d}$$

The solutions of Eqs. (25(c)) and (25(d)) can be represented by the following simple harmonic motion, where ω represents circular frequency:

$$X_1(x_1) = A_1 e_1^{i(\omega/V)x} + A_2 e_1^{-i(\omega/V)x}, \tag{26a}$$

$$T(t) = B_1 e^{i\omega t} + B_2 e^{-i\omega t}. \tag{26b}$$

Comparing Eqs. (23), (24) and (26), we can write the generalized solution of $u(x_1, t)$ as

$$u_1(x_1, t) = C_1 e^{i\omega(t+x_1/V)} + C_2 e^{i\omega(t-x_1/V)} + C_3 e^{-i\omega(t+x_1/V)} + C_4 e^{-i\omega(t-x_1/V)}, \tag{27}$$

where C_1, C_2, C_3, and C_4 represent amplitudes of the wave, and the power of the exponential terms represent corresponding phases.

4.2. *Wentzel, Kramer, Brioullion, and Jeffrey's (WKBJ) Solution of Wave Equation for Heterogeneous Medium*

For a homogeneous medium, the wave equation may be written as

$$\nabla^2 \in = \frac{1}{V^2}\frac{\partial^2 \in}{\partial t^2}, \tag{28}$$

where $\in = \Phi$ for the P wave and $\in = \Psi$ for the S wave, and the corresponding values of velocity V are α and β. For a heterogeneous medium, the wave equation will be much more complex. However, if V varies smoothly in space and has a small gradient, then we can write the wave equation approximately in the same format as in Eq. (28), where $V = V(x)$ with x representing the position of a point in the medium. In that case, a solution of the wave equation can be obtained, although it will be an approximate solution:

$$\nabla^2 \in = \frac{1}{V(x)^2}\frac{\partial^2 \in}{\partial t^2}. \tag{29}$$

Let a plane harmonic wave propagate through the medium for which

$$\in = A(x) e^{i(+\omega t + k \cdot x)}, \tag{30}$$

where $A(x)$ represents wave amplitude at the position $x = (x_1, x_2, x_3)$ and k is the wavenumber representing the direction of wave propagation, i.e., a ray path. Here, $k = \omega/V(x)$. For a homogeneous material,

V and k are constants; hence, k would represent a straight line. For a heterogeneous material, $V = V(x)$, and the ray path will not be a straight line. Let $\omega W(x)/V_0$ replace $k \cdot x$ (where, V_0 is a reference velocity). Then,

$$\in = A(x)e^{i\omega\left(\frac{W(x)}{V_0}-t\right)}. \tag{31}$$

Putting this in Eq. (29), we get

$$\nabla^2\left[A(x)e^{i\omega\left(\frac{W(x)}{V_0}-t\right)}\right] = \frac{1}{V(x)^2}\frac{\partial^2 A(x)e^{i\omega\left(\frac{W(x)}{V_0}-t\right)}}{\partial t^2}. \tag{32}$$

The first term of the left-hand side of Eq. (32) is

$$\frac{\partial^2 \in}{\partial x_1^2} = \frac{\partial}{\partial x}\left[\frac{\partial A(x)e^{i\omega\left(\frac{W(x)}{V_0}-t\right)}}{\partial x_1} + A(x)i\omega\frac{\partial W(x)e^{i\omega\left(\frac{W(x)}{V_0}-t\right)}}{\partial x_1}\right]$$

$$\times e^{i\omega\left(\frac{W(x)}{V_0}-t\right)} = \left[\frac{\partial^2 A(x)}{\partial x_1^2} - \frac{\omega^2 A(x)}{V_0^2}\left(\frac{\partial W(x)}{\partial x_1}\right)^2\right.$$

$$\left.+\,i\left(\frac{2\omega}{V_0}\frac{\partial A(x)}{\partial x_1}\frac{\partial W(x)}{\partial x_1} + A(x)\frac{\omega}{V_0}\frac{\partial^2 W(x)}{\partial x_1^2}\right)\right]. \tag{33}$$

Similarly, $\frac{\partial^2 \in}{\partial x_2^2}$, $\frac{\partial^2 \in}{\partial x_3^2}$, and $\frac{\partial^2 \in}{\partial t^2}$ can be calculated. Putting these values in Eq. (32) and equating real and imaginary parts to zero separately, we get

$$\nabla^2 A(x) - A(x)\frac{\omega^2}{V_0^2}\left[\left(\frac{\partial W(x)}{\partial x_1}\right)^2 + \left(\frac{\partial W(x)}{\partial x_2}\right)^2\right.$$

$$\left.+\left(\frac{\partial W(x)}{\partial x_3}\right)^2\right] = -\frac{\omega^2}{V(x)^2}A(x), \tag{34}$$

$$2\left(\frac{\partial W(x)}{\partial x_1}\frac{\partial A(x)}{\partial x_1} + \frac{\partial W(x)}{\partial x_2}\frac{\partial A(x)}{\partial x_2} + \frac{\partial W(x)}{\partial x_3}\frac{\partial A(x)}{\partial x_3}\right)$$

$$+\,A(x)\nabla^2 W(x). \tag{35}$$

From Eq. (34), we get

$$\left(\frac{\partial W(x)}{\partial x_1}\right)^2 + \left(\frac{\partial W(x)}{\partial x_2}\right)^2 + \left(\frac{\partial W(x)}{\partial x_3}\right)^2 - \frac{V_0^2}{V(x)^2}$$
$$= \frac{V_0^2}{A(x)\omega^2}(\nabla^2 A(x)). \tag{36}$$

For high frequency and small variation in $A(x)$, caused by small variation in $V(x)$, the right-hand side of Eq. (36) turns negligibly small ($\simeq$0). In that case, Eq. (36) can be written as

$$\left(\frac{\partial W(x)}{\partial x_1}\right)^2 + \left(\frac{\partial W(x)}{\partial x_2}\right)^2 + \left(\frac{\partial W(x)}{\partial x_3}\right)^2 = \frac{V_0^2}{V(x)^2} \tag{37}$$

This equation is called Eikonal equation. Solution of this is not an exact representation of the solution of the wave equation. However, for many regions of the Earth, the necessary restrictions on spatial variations of seismic wave velocity are satisfied; hence, the solution is useful. The Eikonal equations are partial-differential equations that relate raypaths to the seismic velocity distribution. The conditions for the geometric ray theory to be a useful approximation of wave equations is $\nabla A(x)$ over one wavelength (λ) must be smaller than $A(x)$. Let reference wavelength $\lambda_0 = 2\pi V_0/\omega$. For the condition, for Eq. (37) to hold, we require that (from Eq. (36))

$$\lambda_0^2(\nabla^2 A(x)/A(x)) \ll \nabla W(x) \cdot \nabla W(x). \tag{38}$$

This gives

$$\lambda_0^2(\nabla^2 A(x)/A(x)) \ll V_0^2/V^2(x). \tag{39}$$

For weak heterogeneity, $V_0^2/V^2(x) - 1$. Therefore, $\lambda_0^2(\nabla^2 A(x)/A(x)) \ll 1$. To gain physical insight from Eq. (37), we write it as follows:

$$\nabla W(x) \cdot \nabla W(x) = V_0^2/V^2(x) \tag{40a}$$
$$\text{or} \quad \nabla W(x) \simeq V_0/V(x). \tag{40b}$$

On the other hand, from Eq. (35) we can write

$$\nabla^2 W(x) = -2\nabla W(x) \cdot \nabla A(x)/A(x) \tag{41}$$

$$\frac{\nabla A(x)}{A(x)} = -\frac{1}{2}\frac{\nabla^2 W(x)}{\nabla W(x)} = -\frac{1}{2}\frac{\nabla\left(\frac{V_0}{V(x)}\right)}{\left(\frac{V_0}{V(x)}\right)} = \frac{1}{2}\frac{\nabla V(x)}{V(x)}. \tag{42}$$

5. Derivation of Raypath Geometry from Eikonal Equation

In this section onwards, we will consider vertically downward direction as positive x_3 direction. This is because the variation of Earth's property is commonly estimated with depth in Earth Sciences. Figure 2 shows a schematic representation of a wavefront and the corresponding raypath. The latter is the direction of propagation of the former, and hence, it is always perpendicular to the wavefront at any given location and time. It is to be noted that the wavenumber k is related to the direction of propagation, and in the previous equations, we have introduced a parameter $W(x)$ such that $\omega W(x)/V_0 = k \cdot x$.

In Figure 2, the direction cosines for the raypath ds can be given as dx_1/ds, dx_2/ds, and dx_3/ds, and they satisfy the following relation:

$$\left(\frac{dx_1}{ds}\right)^2 + \left(\frac{dx_2}{ds}\right)^2 + \left(\frac{dx_3}{ds}\right)^2 = 1. \tag{43}$$

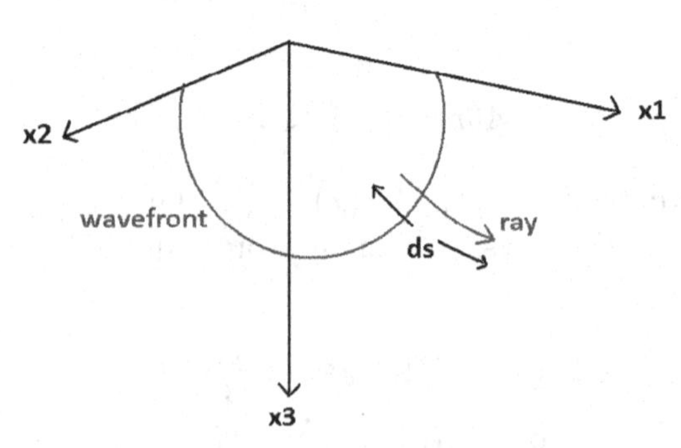

Figure 2. Schematic representation of a wavefront in 3D. A raypath of length ds is also shown [2].

In Eq. (31), the power of the exponential term represents the phase of the wave. From the definition of a wavefront, this term is a constant for a given frequency and time over a given wavefront. This means that the term $\omega W(x)/V_0$ is constant over a wavefront. As V_0 is a constant, then for a given ω, $W(x)$ is constant over a given wavefront and can be considered as a parameter representing it. This means that its gradient $\nabla W(x)$ will be nonzero only in a direction perpendicular to the wavefront, which is the direction of a raypath along which the wave will propagate. Thus, we can say that $\partial W(x)/\partial x_i$ is proportional to dx_i/ds. Let the constant of proportionality be a. Then Eq. (43) can be written as

$$a\left(\frac{\partial W(x)}{\partial x_1}\right)^2 + a\left(\frac{\partial W(x)}{\partial x_2}\right)^2 + a\left(\frac{\partial W(x)}{\partial x_3}\right)^2 = 1, \tag{44}$$

where $a^{-1} = n = \frac{V_0^2}{V(x)^2}$ is the index of refraction. Combining Eqs. (43) and (44), we get the following equations, termed as normal equations:

$$n\frac{dx_1}{ds} = \frac{\partial W(x)}{\partial x_1}, \tag{45a}$$

$$n\frac{dx_2}{ds} = \frac{\partial W(x)}{\partial x_2}, \tag{45b}$$

$$n\frac{dx_3}{ds} = \frac{\partial W(x)}{\partial 3}. \tag{45c}$$

The nature of variation of normal equations along a raypath leads us to the raypath equation. Taking derivative of the normal equation with respect to ds, we get

$$\begin{aligned}\frac{d}{ds}\left(n\frac{dx_1}{ds}\right) &= \frac{d}{ds}\left(\frac{\partial W(x)}{\partial x_1}\right)\\ &= \frac{\partial}{\partial x_1}\left(\frac{\partial W(x)}{\partial x_1}\frac{dx_1}{ds} + \frac{\partial W(x)}{\partial x_2}\frac{dx_2}{ds} + \frac{\partial W(x)}{\partial x_3}\frac{dx_3}{ds}\right)\\ &= \frac{\partial}{\partial x_1}\left[n\left\{\left(\frac{dx_1}{ds}\right)^2 + \left(\frac{dx_2}{ds}\right)^2 + \left(\frac{dx_3}{ds}\right)^2\right\}\right] = \frac{\partial n}{\partial x_1}.\end{aligned} \tag{46}$$

The generalized form of this equation is

$$\frac{d}{ds}\left(n\frac{dx_i}{ds}\right) = \frac{\partial n}{\partial x_i} \tag{47a}$$

$$\text{or} \quad \frac{d}{ds}\left(\frac{1}{V(x)}\frac{dx}{ds}\right) = \nabla\left(\frac{1}{V(x)}\right). \tag{47b}$$

This second-order differential equation represents a raypath that depends on the spatial rate of change of velocity. The following two initial conditions decide the direction in which a ray would turn in a given medium: (i) direction in which the ray leaves some arbitrary reference point $(\partial x/\partial s)|_{s=0}$; (ii) location of the reference point s_0 in the medium. To explain this, a simple case is considered, where $V(x) = V(x_3)$, i.e., the velocity varies only in the vertical direction. In such a case, we can write $n = n(x_3)$, $\partial n/\partial x_1 = \partial n/\partial x_2 = 0$. In this case, we can write

$$\left(n\frac{dx_1}{ds}\right) = V_1 \text{ (let)} = \text{constant}, \tag{48a}$$

$$\left(n\frac{dx_2}{ds}\right) = V_2 \text{ (let)} = \text{constant}, \tag{48b}$$

$$\text{and} \quad \frac{d}{ds}\left(n\frac{dx_3}{ds}\right) = \frac{\partial n}{\partial x_3}. \tag{48c}$$

The ratio of V_1/V_2 would confine the raypath to a plane perpendicular to the x_1x_2 plane, so its projection into the x_1x_2 plane would be a straight line. Let the coordinate be rotated such that the x_1 axis lies along the line along which the projection of raypath crosses the x_1x_2 plane (Figure 3(a)). Let at an arbitrary point the raypath makes an angle of i with x_3 axis. Let the variation of velocity with depth is as given in Figure 3(b). From Figure 3(a), we may write

$$\frac{dx_1}{ds} = \sin i \tag{49a}$$

$$\text{and} \quad \frac{dx_3}{ds} = \cos i. \tag{49b}$$

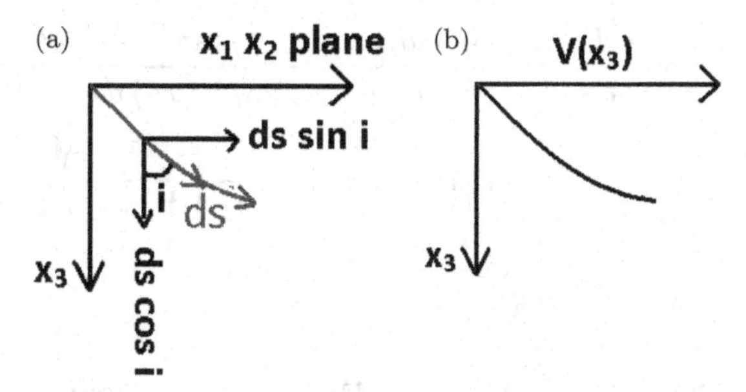

Figure 3. (a) Raypath in a vertically heterogeneous medium and (b) variation of velocity with depth [2].

Therefore, from Eqs. (48(a)) and (49(a)), we can write

$$n\frac{dx_1}{ds} = \frac{V_0}{V(x)}\sin i = \mathit{constant} \tag{50a}$$

$$\text{or} \quad \frac{\sin i}{V(x)} = p = \mathit{constant}. \tag{50b}$$

Equation (50(b)) is nothing but Snell's law, where p is called the ray parameter or horizontal slowness, and slowness is the inverse of velocity. This equation states that for a given raypath, horizontal slowness is constant. When $p = 0$, wave travels in the x_3 direction, and when $p = 1/V$, wave travels in the horizontal direction.

From Eq. (48(c)),

$$\begin{aligned}\frac{dn}{dx_3} &= \frac{d}{ds}\left(n\frac{dx_3}{ds}\right)\frac{d(n\cos i)}{ds}\\ &= \frac{dn}{ds}\cos i + n\frac{d\cos i}{ds} = \cos i\frac{dn}{dx_3}\frac{dx_3}{ds} + n\frac{d\cos i}{di}\frac{di}{ds}\\ &= -n\sin i\frac{di}{ds} + \cos^2 i\frac{dn}{dx_3},\end{aligned} \tag{51a}$$

$$\text{or} \quad (1-\cos^2 i)\frac{dn}{dx_3} = -n\sin i\frac{di}{ds}, \tag{51b}$$

$$\text{or} \quad \sin^2 i\frac{dn}{dx_3} = -n\sin i\frac{di}{ds}, \tag{51c}$$

$$\text{or} \quad \frac{di}{ds} = -\frac{\sin i}{n}\frac{dn}{dx_3} = -\frac{\sin i}{V_0/V(x_3)}\frac{d}{dx_3}\left(\frac{V_0}{V(x_3)}\right)$$
$$= -\sin i \cdot V(x_3) \cdot \frac{d}{dx_3}\left(\frac{1}{V(x_3)}\right)$$
$$= \frac{\sin i}{V(x_3)}\frac{dV(x_3)}{dx_3}, \tag{51d}$$

where $n = V_0/V(x_3)$. Finally, from Eq. (51(c)), we get

$$\frac{di}{ds} = p\frac{dV(x_3)}{dx_3}. \tag{52}$$

Equation (52) shows that the curvature of the raypath is proportional to the velocity gradient. Figure 4 shows an example of how the curvature of raypath depends on the velocity gradient when velocity varies linearly with depth. When velocity increases with depth, the raypath curves upward, and when the velocity decreases with depth, it curves downward. The corresponding wavefront positions at successive times are also shown. When the velocity is larger, the wavefronts are further away. Hence, in Figure 4, in the left panel, where velocity increases with depth, the wavefronts are further apart in the deeper region and in the right panel, the opposite is true.

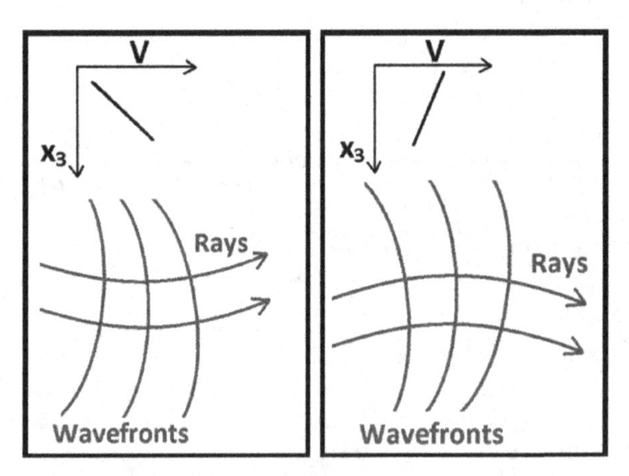

Figure 4. Schematic representation of the nature of wavefront and the corresponding raypaths for different velocity models [2].

From Eq. (48(c)), several important conclusions can be drawn. They are as follows:

(i) For each angle i, a specific ray leaves the source and follows a specific raypath.
(ii) The initial angle and the velocity structure determine the distance at which the ray will emerge at the surface.
(iii) For a given source–receiver geometry, several possible connecting raypaths may exist, which means that a multiplicity of arrival will occur, all with different initial angles and travel times. This is the basis for seismic interpretation of Earth's structure.

6. Derivation of Relation Between Travel Time and Epicentral Distance for Source at Surface When Velocity Increases with Depth

From Figure 5, we can write

$$\sin i = \frac{dx_1}{ds} = pV[V = V(x_3)], \tag{53a}$$

$$\cos i = \frac{dx_3}{ds} = \sqrt{1 - \sin^2 i} = \sqrt{1 - p^2V^2}. \tag{53b}$$

Therefore,

$$dx_1 = ds \sin i = \frac{dx_3}{\cos i} pV = \frac{pV}{\sqrt{1 - p^2V^2}} dx_3 \tag{54}$$

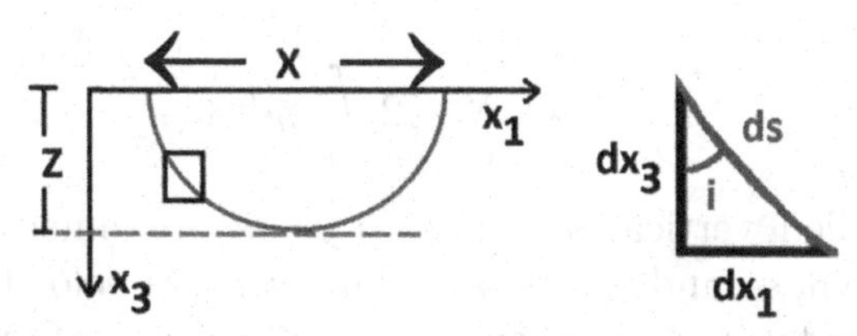

Figure 5. Left panel represents a raypath in a medium where the velocity increases with depth. Right panel represents a small part of the same raypath present within the boxed portion shown in the left panel. z and i represent the maximum depth of penetration of the raypath and the angle the raypath makes at an arbitrary location with the vertical axis, respectively [2].

Equation (54(c)) shows that the horizontal distance covered by a raypath depends on its ray parameter p. For a given raypath, it is constant. Hence, each raypath will have a unique ray parameter. By integrating Eq. (54(c)), we obtain the epicentral distance X, which we choose to write as a function of p:

$$X(p) = 2\int_0^z \frac{pV}{\sqrt{1-p^2V^2}}\,dx_3, \tag{55a}$$

which can also be written as

$$X(p) = 2p^2\int_0^z \frac{dx_3}{\sqrt{\gamma^2-p^2}}, \tag{55b}$$

where $\gamma = \dfrac{1}{V(x_3)}$.

The travel time along the small slice of raypath ds is given by

$$dT = \frac{ds}{V}. \tag{56a}$$

Hence, the travel time T along the raypath is given by

$$T = 2\int_{Path} \frac{ds}{V} = 2\int_0^z \frac{dx_3}{V\cos i}, \tag{56b}$$

$$\text{or}\quad T = 2\int_0^z \frac{\gamma^2 dx_3}{\sqrt{\gamma^2-p^2}} = 2\int_0^z \left(\frac{p^2}{\sqrt{\gamma^2-p^2}} + \sqrt{\gamma^2-p^2}\right)dx_3$$

$$= pX + 2\int_0^z \sqrt{\gamma^2-p^2}dx_3, \tag{56c}$$

$$\text{or}\quad T = pX + 2\int_0^z \eta dx_3, \tag{56d}$$

where η is called vertical slowness, p or ray parameter being the horizontal slowness, and slowness is the inverse of velocity. Equation (56(d)) is called travel time equation. It consists of two separate terms: one depends on the horizontal distance or epicentral distance and the other depends on the distance in the vertical direction or depth. It relates travel time of a seismic wave from source to receiver along a raypath with epicentral distance. These two parameters can

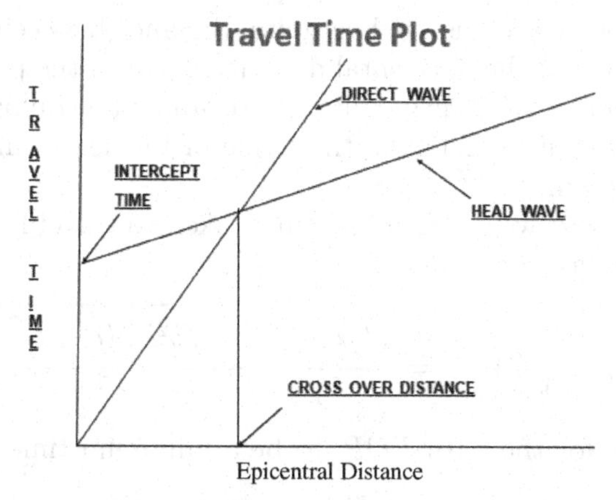

Figure 6. Schematic representation of travel time plot for direct and refracted waves for source at surface.

be estimated from observed records of earthquakes, i.e., seismograms. A plot of T versus X is called a travel time curve (Figure 6).

From Eq. (56(d)), we can obtain

$$\frac{dT}{dX} = p = \frac{\sin i}{V(x_3)} \tag{57}$$

The left-hand side of the equation can be estimated from observed data. This equation is used extensively in seismology for estimation of seismic wave velocity variation with depth within the Earth. It is to be noted that the rate of change of seismic wave velocity is much higher with x_3 compared to that with respect to x_1 or x_2, i.e., it changes more rapidly with depth compared to horizontal direction.

7. Interpretation of Snell's Law Using Simple Ray Geometry

In this section, we shall explain the Snell's law using simple ray geometry and Fermat's principle of minimum travel time of a wave from one point to another. Let a raypath start from a point P in a medium with velocity V_1 and cross a boundary at point O to reach a point P' in the next medium with velocity V_2, where d and e are the path lengths of the ray in the first and second media, respectively, a and b are the vertical distances of the points P and P' from the interface, respectively, c is the horizontal distance between the points P and P', and X is the horizontal distance between the points P and O. As the velocity in each medium is constant, the raypath in each one is a straight line. Let i be the angle of incidence and r be the angle of refraction.

From Figure 7, we see that the time taken to travel from P to P' can be written as

$$T_{pp'} = \frac{d}{V_1} + \frac{e}{V_2} = \frac{\sqrt{a^2 + x^2}}{V_1} + \frac{\sqrt{b^2 + (c - x)^2}}{V_2}. \tag{58}$$

The condition for the path POP' to be a minimum time path is

$$\frac{dT}{dX} = 0. \tag{59}$$

Applying this condition on Eq. (58), we get

$$\frac{x}{V_1\sqrt{a^2 + x^2}} - \frac{c - x}{V_2\sqrt{b^2 + (c - x)^2}} = 0 \tag{60}$$

$$\text{or} \quad \frac{\sin i}{V_1} = \frac{\sin r}{V_2} = p. \tag{61}$$

This is the classical Snell's law that states that the ray parameter p, which represents the horizontal component of slowness along a raypath, remain constant even when the raypath changes direction because of variation of medium velocity. Even with continuous change in velocity, the ray parameter along a given raypath remains constant (Eq. (50(b))).

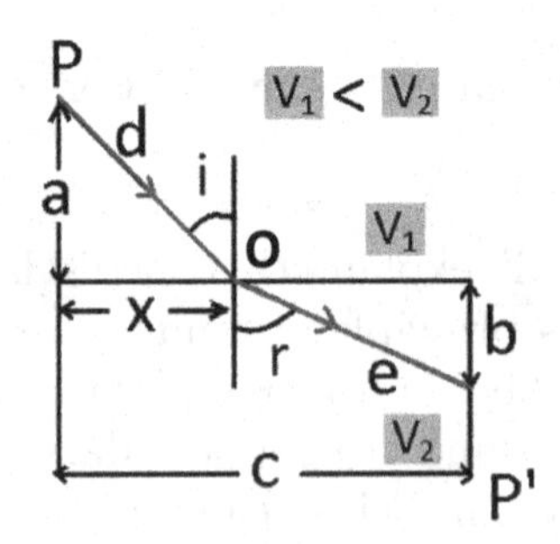

Figure 7. Refraction of a raypath crossing a boundary between two homogeneous media.

8. Travel Time Equations

8.1. *Travel Time Equations in a Layered Earth*

When seismic waves are recorded within a few hundred kilometers of the source, they can be used to study the velocity variation of the upper layer called crust and the top of the underlying medium called mantle. In such situations normally, waves recorded at a recording station may travel directly from the earthquake source to the station (called direct wave), get reflected from the crust–mantle boundary (called reflected wave), or may get refracted along the same boundary (called refracted or head waves). In such cases, Cartesian coordinates may be used for calculation of travel time of each type. Let the layer have a constant velocity V_1 and thickness H and the underlying medium (here assumed to be a half space) have a velocity V_2. When the interface between the two media is horizontal (Figure 8) and both the source and station are at the surface, the travel time equations for the respective types may be given as, for a direct wave,

$$t_d = \frac{x}{V_1}, \tag{62}$$

and for a reflected wave,

$$t_r = \frac{AB + BC}{V_1} = \frac{\sqrt{x^2 + 4H^2}}{V_1}. \tag{63}$$

If $V_1 < V_2$ then the angle of refraction r is greater than the angle of incidence i. At critical angle of refraction, i.e., when $r = 90°$, the angle of incidence i_c is given by (from Eq. (61))

$$i_c = \sin^{-1}(V_1/V_2). \tag{64}$$

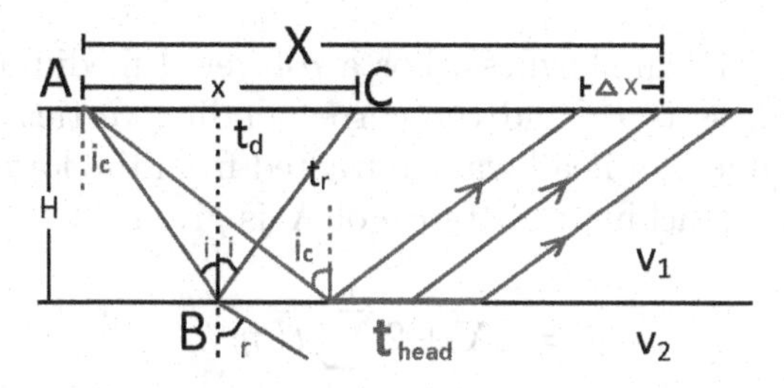

Figure 8. Schematic diagram of raypaths in a layer over half-space medium.

The horizontal distance at which this ray reaches the surface is given by

$$X_{critical} = 2H \tan(i_c), \tag{65}$$

and the travel time along this raypath is given by

$$t_{critical} = \frac{\sqrt{(2H \tan i_c{}^2 + 4H^2}}{V_1} = \frac{2H}{V_1 \cos i_c}. \tag{66}$$

The refracted wave will travel along the interface with a velocity of V_2. According to Fermat's principle, each point this raypath reaches will act as a secondary source and will distribute the resultant energy along a wavefront that will travel in all directions. However, the minimum time path for this wave will follow Snell's law. As a consequence, the corresponding raypaths reaching the surface will be as shown parallel to each other in Figure 8. The distance where this wave has reached can be given as

$$X_{head} = 2H \tan i_c + \Delta x, \tag{67}$$

where Δx is the distance traveled by the wave along the interface with a velocity of V_2. Corresponding travel time is as follows:

$$\begin{aligned} T_{head} &= \frac{2H}{V_1 \cos i_c} + \frac{\Delta x}{V_2} = \frac{2H}{V_1 \cos i_c} + \frac{X_{head} - 2\tan i_c}{V_2} \\ &= \frac{X_{head}}{V_2} + \frac{2H}{V_1 \cos i_c}\left(1 - \frac{V_1 \sin i_c}{V_2}\right) = \frac{X_{head}}{V_2} + \frac{2H}{V_1 \cos i_c} \\ &\quad \times (1 - \sin^2 i_c) = \frac{X_{head}}{V_2} + \frac{2H \cos i_c}{V_1} = pX_{head} + 2H\eta. \end{aligned} \tag{68}$$

Here, η is the vertical slowness. For a n-layered medium, where the ith layer thickness is H_i and the corresponding vertical slowness is η_i, the travel time of a head wave refracted from the lower boundary of the nth layer reaching a distance of X is given by

$$T = pX + 2\sum_{i=1}^{n} H_i \eta_i. \tag{69}$$

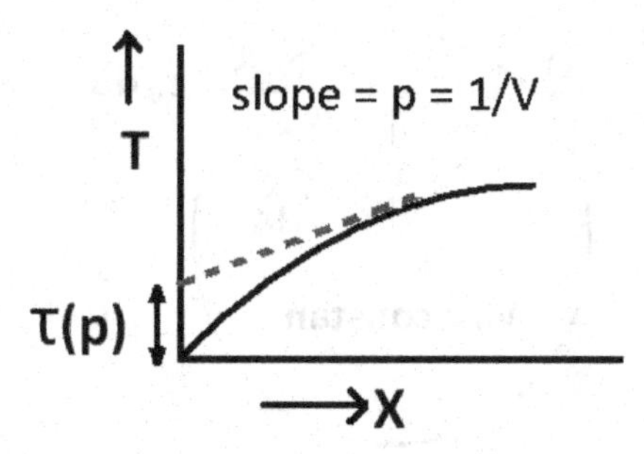

Figure 9. Travel time curve for a medium where velocity increases continuously with depth [2].

For a medium where the velocity continuously increases with depth, the corresponding travel time equation will be (Figure 9)

$$T = pX + 2\int_0^{zmax} \eta dz = pX + 2\int_0^{zmax} \sqrt{\gamma^2 - p^2}dz, \qquad (70)$$

where γ is as given in Eq. (55(b)). A line tangent to this travel time curve (Figure 9) at a point where epicentral distance is X crosses the T axis at a time called intercept time $\tau(p)$, where

$$\tau(p) = T - pX = 2\int_0^{z} \sqrt{\gamma^2 - p^2}dx_3 \qquad (71)$$

and

$$\frac{d\tau}{dp} = \frac{d}{dp}(2\int_0^{z} \sqrt{\gamma^2 - p^2}dx_3) = 2\int_0^{z} -\frac{p}{\sqrt{\gamma^2 - p^2}}dx_3 = -X. \quad (72)$$

Equations (71) and (72) are routinely used in Seismology to estimate velocity variation with depth in the Earth.

8.2. *Travel Time Equations in a Sphere*

For a spherical medium, the epicentral distance is given in terms of angular distance Δ, where Δ is the angle subtended by two lines: one connecting the source to the centre of the sphere, the other connecting the station to its centre. For a homogeneous sphere (Figure 10), the raypaths (e.g., OA) will be straight lines. Let r_0 be the radius

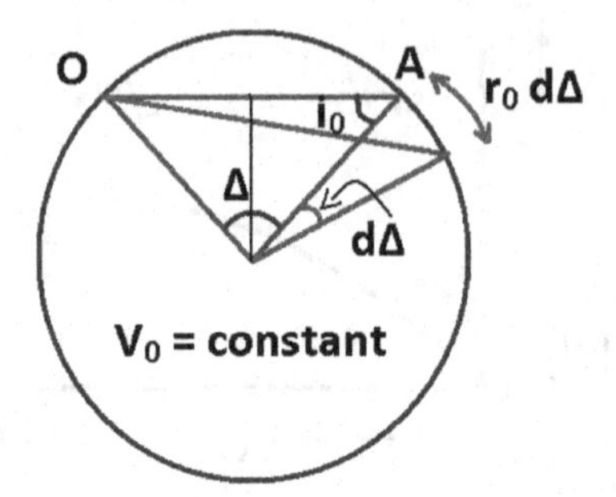

Figure 10. Representation of raypath in a homogeneous sphere and the corresponding epicentral distance [2].

and V_0 be the velocity of seismic wave for the sphere. Travel time along the raypath from O to A will be given by

$$T(\Delta) = \frac{OA}{V} = \frac{2r_0 \sin\left(\frac{\Delta}{2}\right)}{V_0}, \tag{73}$$

where $i_0 = 90^\circ - \frac{\Delta}{2}$ is the angle of incidence and the ray parameter p is given by

$$p = \frac{r_0 \sin(i_0)}{V_0} = \frac{r_0 \cos\left(\frac{\Delta}{2}\right)}{V_0}. \tag{78}$$

It is to be noted that unlike the horizontal layer case, the travel time equation for direct arrival is not an equation of a straight line. Also, the ray parameter value decreases with increasing epicentral distance.

8.3. *Ray Parameter for a Spherical Earth*

When we deal with earthquake waves that have traveled more than 1000 km, then the spherical nature of the Earth cannot be ignored. To take care of such situations, in this section, we derive travel time equations for a sphere made up of concentric shells of different velocities and thicknesses (Figure 11). Let a raypath fall on one boundary at point P at an angle θ_1, where the angle is measured with respect

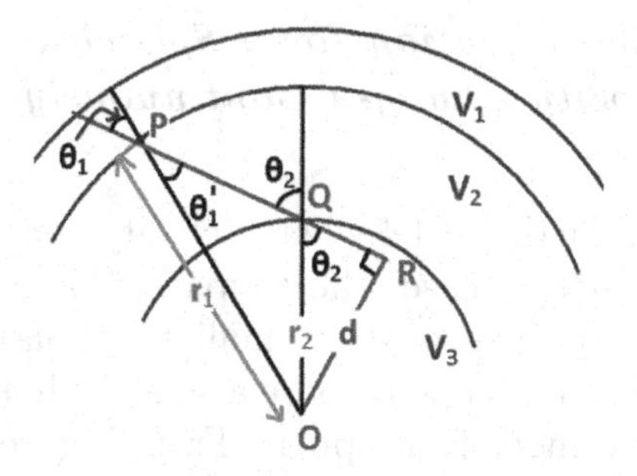

Figure 11. Part of a sphere made up of concentric shells of different velocities [2].

to the radius of the sphere. It gets refracted at an angle θ_1' into the second medium. According to Snell's law,

$$\frac{\sin\theta_1}{V_1} = \frac{\sin\theta_1'}{V_2}. \tag{79}$$

Let the raypath falls on the second boundary at Q at an angle of θ_2. Let us extend the straight line PQ up to a point R and join it to the centre of the sphere O. Let $OR = d$. In this case, we can write

$$d = r_1 \sin\theta_1' = r_2 \sin\theta_2, \tag{80}$$

where r_1 and r_2 are the radii of the spherical shells. Putting this in Eq. (79), we get

$$\frac{r_1 \sin\theta_1}{V_1} = \frac{r_1 \sin\theta_1'}{V_2} = \frac{r_2 \sin\theta_1}{V_2}. \tag{81}$$

Since r_1 and r_2 can have any arbitrary values along the raypath, this is a general equation along the entire raypath. In other words, the ray parameter p for a spherical body may be written as

$$\frac{r \sin i}{V} = p. \tag{82}$$

It is to be noted that the dimension of ray parameter for a spherical representation of the earth is different from that for a flat representation. However, the meaning of ray parameter is same for both the cases.

8.4. Travel Time Equation for a Spherical Body where Velocity Changes Continuously with Depth

Let us consider an Earth model where velocity continuously changes along the radius. In this case, the raypath will continuously bend (Figure 12(a)). If we take a very small part (ds) of the raypath (Figure 12(b)), we can assume it to be a straight line. Let ds subtend an angle $d\Delta$ at the centre of the sphere. From Figure 12, we can write

$$(ds)^2 = (dr)^2 + r^2(d\Delta)^2 \tag{83}$$

and

$$\sin i = r\frac{d\Delta}{ds}. \tag{84}$$

Therefore,

$$p = \frac{r \sin i}{V} = \frac{r^2}{V}\frac{d\Delta}{ds} \tag{85}$$

and

$$ds = (\frac{1}{p})r^2\frac{d\Delta}{V}. \tag{86}$$

This gives

$$\frac{1}{p^2}\frac{r^4}{V^2}(d\Delta)^2 - r^2(d\Delta)^2 = (dr)^2 \tag{87}$$

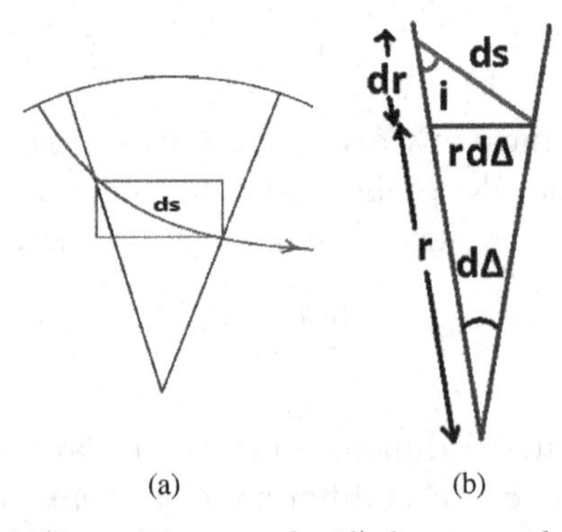

Figure 12. Raypath (line with arrow head) for an earth model where velocity varies continuously with depth.

or

$$d\Delta = \frac{pV}{r}\frac{dr}{\sqrt{r^2 - p^2V^2}} = \frac{pdr}{r\sqrt{\varepsilon^2 - p^2}}, \tag{88}$$

where $\varepsilon = r/V$. Hence, epicentral distance may be given as

$$\Delta = 2p\int_{r_i}^{r_0} \frac{dr}{r\sqrt{\varepsilon^2 - p^2}}. \tag{89}$$

Again,

$$(ds)^2 = (dr)^2 + r^2\left(\frac{pVds}{r^2}\right)^2 \tag{90}$$

or

$$ds = \frac{dr}{\sqrt{1 - \left(\frac{p^2V^2}{r^2}\right)}} = \frac{\varepsilon dr}{\sqrt{\varepsilon^2 - p^2}}. \tag{91}$$

Therefore, the travel time T along the path is given as (assuming source is at the surface)

$$\begin{aligned} T &= 2\int_{Path} \frac{ds}{V} = 2\int_{r_t}^{r_0} \frac{\varepsilon dr}{V\sqrt{\varepsilon^2 - p^2}} = 2\int_{r_t}^{r_0} \frac{\varepsilon^2 dr}{r\sqrt{\varepsilon^2 - p^2}} \\ &= p\Delta + 2\int_{r_t}^{r_0} \frac{\sqrt{\varepsilon^2 - p^2}dr}{r} \end{aligned} \tag{92a}$$

or

$$T = p\Delta + 2\int_{r_t}^{r_0} \frac{\sqrt{\varepsilon^2 - p^2}dr}{r}. \tag{92b}$$

The first term depends on the horizontal (epicentral) distance and the second term depends on the vertical (radial) distance. This is similar to the case where the Earth was modeled as a flat medium. This shows that irrespective of the coordinate system used, the travel time equation has the same type of representation. The second term in Eq. (92(b)) is the $\tau(p)$ term similar to that of the flat Earth model given in Eq. (71).

9. Amplitude of Seismic Wave

The energy released by a seismic source propagates through the medium as a seismic wave. The net energy gets distributed over the wavefront. As the wave propagates further and further, the surface area of the wavefront increases. Hence, the energy per unit area of the wavefront decreases. This is called geometrical spreading. The amplitude of a seismic wave recorded at any location at any time will be proportional to the square root of energy per unit area of the wavefront passing through that point at that time. In this section, how amplitude gets affected when the wave propagates through a medium where velocity changes continuously and where there is a sharp jump in velocity will be discussed. In all of the discussion here, we assume that the medium is elastic, hence the amplitude variation caused by inelastic attenuation is ignored.

9.1. *Amplitude of Seismic Wave Recorded at Surface for a Medium with Continuous Velocity Variation*

In this section, certain assumptions are made for ease of derivation. The insight thus gleaned throws light upon how amplitude of seismic wave varies. In the following derivations, it is assumed that the wavefront is hemispherical, although when velocity changes with position, it will not be. However, the inferences drawn will still be valid to a large extent. We assume that the rate of change of velocity is rather small, hence the wavefront will be almost hemispherical. Let there be a point source (centre of hemispherical shell in Figure 13(a)).

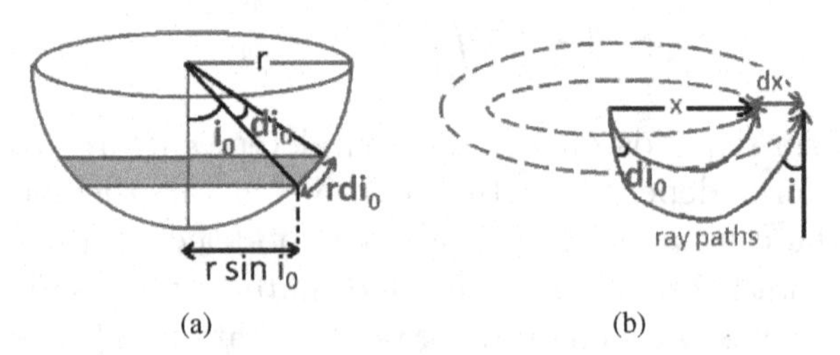

Figure 13. (a) Hemispherical wavefront at a small distance from a point source and (b) Raypaths at small angular distance di_0 apart.

The hemisphere of radius r represents the wavefront. We take a hemispherical wavefront because the energy transfer from seismic wave propagating through the earth to air is negligibly small. For all practical purposes, air is considered as vacuum when seismic wave propagation problems are considered. Let the total energy released from the source be K. Then energy per unit area of the wavefront is

$$\frac{\text{Energy}}{\text{Area}} = \frac{K}{2\pi r^2}. \tag{93a}$$

Let a bundle of rays leave the source between angle i_0 and $i_0 + di_0$ (Figure 13). The energy E carried by this bundle would pass through the shaded strip on the wavefront (Figure 13(a)), where

$$E = \left(\frac{K}{2\pi r^2}\right)(2\pi r \sin i_0)(di_0 r) = K \sin i_0\, di_0. \tag{93b}$$

The area on the surface of the Earth (a circular strip of width dx at a distance x; Figure 13(b) where this energy will reach is

$$A = 2\pi x dx \cos i_0. \tag{93c}$$

Hence, the energy per unit area on the Earth's surface will be

$$E(x) = \left(\frac{K}{2\pi}\right)\left(\frac{\tan i_0}{x}\right)\left(\frac{di_0}{x}\right). \tag{93d}$$

Now, $p = \sin i_0/V_0 = dT/dx$ (from Eq. (57)) giving us

$$i_0 = \sin^{-1}\left(V_0 \frac{dT}{dx}\right). \tag{93e}$$

This gives us

$$\frac{di_0}{dx} = \frac{V_0}{\sqrt{1 - V_0^2\left(\frac{dT}{dx}\right)^2}}\left(\frac{dT}{dx}\right)^2 = \frac{V_0}{\cos i_0}\frac{d^2T}{dx^2}. \tag{93f}$$

Therefore,

$$E(x) = \left(\frac{K}{2\pi}\right) V_0 \left(\frac{\tan i_0}{\cos i_0 x}\right)\frac{d^2T}{dx^2} = \left(\frac{K}{2\pi}\right)\frac{V_0^2 p}{(1 - p^2V_0^2)}\frac{dp}{dx}. \tag{93g}$$

From this, it can be inferred that the energy received per unit area on the surface will be proportional to the total energy released. Also, the energy per unit area will vary rapidly when the rate of change of ray parameter with epicentral distance (dp/dx) is large. The last parameter has a larger value when the velocity changes more rapidly with depth. As the amplitude of seismic wave is proportional to the square root of energy per unit area the seismic records above two inferences can be verified from observational data. It is reported that observational data supports these inferences.

9.2. *Energy Partitioning Caused by Presence of an Interface Across where Velocity Changes Abruptly*

The Earth is made up of layers with a sudden jump in velocity at their interfaces. Seismic energy falling on such boundaries split up; some get reflected and some get refracted or transmitted to the adjacent layer. If the medium is solid, then mode conversion, *viz.*, P to S or S to P conversion would also occur. Figure 14 shows a schematic representation of what happens to an incident P wave (P_i) when it hits a solid–solid, fluid–solid, fluid–fluid, and solid–vacuum boundary. It is to be noted that the S wave cannot travel through any type of fluid, and no wave can travel through vacuum. In the first case, there will be four off springs, *viz.*, reflected $P(P_r)$, converted and reflected $S(S_r)$, transmitted $P(P_t)$, and converted and transmitted $S(S_t)$. In the second case, there will be P_r, P_t, and S_t, in the third case, there will be P_r and P_t, and in the fourth case, there will be P_r and S_r

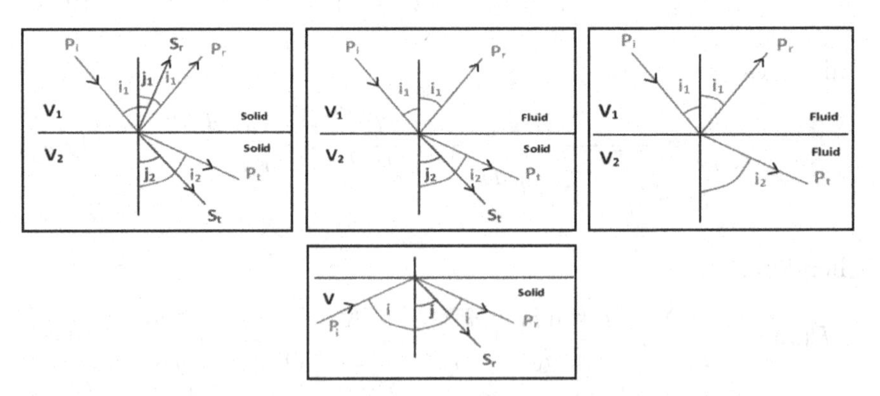

Figure 14. Schematic representation of reflection, transmission, and mode conversion at an interface.

(Figure 14). The angle of incidence, reflection, and transmission are as shown in the figure. It is to be noted that mode conversion can occur only in solid medium. If both the media around the interface are homogeneous and isotropic, then the P wave particle motion will be confined to a vertical plane. Particle motion for the S wave confined to the same vertical plane is called SV wave. P and SV waves can produce converted $S(SV)$ and converted P, respectively, as their particle motion will be confined to the same vertical plane. Let us assume that this vertical plane is the x_1x_3 plane. In that case, the particle motion for the horizontal component of S wave (called SH wave) will be in the x_2 direction. Conversion from SH to P is not possible as the particle motion in these two cases are orthogonal to each other.

The amplitude of such reflected, transmitted, and converted phases would depend on the amplitude of the incident wave, angle of incidence, and the properties of the media on both side of the interface, *viz.*, velocities of P and S waves and densities of the two media. When the amplitudes of such waves are divided by that of the incidence wave, the corresponding parameters are called reflection (R) and transmission (T) coefficients. Hence,

$$R = \frac{\text{Reflected wave amplitude}}{\text{Incident wave amplitude}}, \tag{94a}$$

$$T = \frac{\text{Transmitted wave amplitude}}{\text{Incident wave amplitude}}. \tag{94b}$$

Both R and T can be estimated using certain boundary conditions related to continuity of displacement vector u_i and stress tensor τ_{ij} across the boundary, where both i and j can have three values, *viz.* 1, 2, and 3 and they represent the three axes of the Cartesian coordinates. We assume that x_1 and x_2 are two mutually perpendicular axes in the horizontal plain and x_3 is the vertical axis, with x_3 taken to be positive in the downward direction. We assume that the boundary lies at $x_3 = 0$. In the following notation, the number within brackets represent the medium number and is not the power of the term, and $|x_3 = 0$ represents "at $x_3 = 0$". Depending on the nature of the two media, the boundary conditions are as follows.

(i) For a solid–solid interface, the boundary, conditions are as follows:

(a) Normal component of displacement is continuous across the boundary, i.e.,

$$u_3^{(1)}|_{x3=0} = u_3^{(2)}|_{x3=0}. \tag{95a}$$

(b) Tangential components of displacements are continuous across the boundary, i.e.,

$$u_1^{(1)}|_{x3=0} = u_1^{(2)}|_{x3=0}, \tag{95b}$$

$$u_2^{(1)}|_{x3=0} = u_2^{(2)}|_{x3=0}. \tag{95c}$$

(c) Normal component of stress is continuous across the boundary, i.e.,

$$\tau_{33}^{(1)}|_{x3=0} = \tau_{33}^{(2)}|_{x3=0}. \tag{95d}$$

(d) Tangential components of stress are continuous across the boundary, i.e.,

$$\tau_{13}^{(1)}|_{x3=0} = \tau_{13}^{(2)}|_{x3=0}, \tag{95e}$$

$$\tau_{23}^{(1)}|_{x3=0} = \tau_{23}^{(2)}|_{x3=0}, \tag{95f}$$

$$\tau_{12}^{(1)}|_{x3=0} = \tau_{12}^{(2)}|_{x3=0}. \tag{95g}$$

These conditions will be satisfied only when the two media are at welded contact.

(ii) For a fluid–solid interface, the boundary conditions are as follows:

(a) Normal component of displacement is continuous across the boundary, i.e.,

$$u_3^{(1)}|_{x3=0} = u_3^{(2)}|_{x3=0}. \tag{96a}$$

(b) Normal component of stress is continuous across the boundary, i.e.,

$$\tau_{33}^{(1)}|_{x3=0} = \tau_{33}^{(2)}|_{x3=0}. \tag{96b}$$

(c) Shear components of stress are all zero, i.e.,

$$\begin{aligned}\tau_{13}^{(1)}|_{x3=0} &= \tau_{13}^{(2)}|_{x3=0} = \tau_{23}^{(1)}|_{x3=0} = \tau_{23}^{(2)}|_{x3=0}\\ &= \tau_{12}^{(1)}|_{x3=0} = \tau_{12}^{(2)}|_{x3=0} = 0.\end{aligned} \tag{96c}$$

(iii) For a fluid–fluid interface, the boundary conditions are as follows:

(a) Normal component of displacement is continuous across the boundary, i.e.,

$$u_3^{(1)}|_{x3=0} = u_3^{(2)}|_{x3=0}. \tag{97a}$$

(b) Normal component of stress is continuous across the boundary, i.e.,

$$\tau_{33}^{(1)}|_{x3=0} = \tau_{33}^{(2)}|_{x3=0}. \tag{97b}$$

(iv) For a free surface, i.e., the boundary between solid and vacuum, there is no boundary condition related to displacement as the displacement of the solid medium is totally unconstrained. The available boundary condition is in terms of stress. As stress cannot be transferred to vacuum the conditions are as follows:

$$\tau_{33/x_3=0} = \tau_{13/x_3=0} = \tau_{23/x_3=0} = \tau_{12/x_3=0} = 0. \tag{98a}$$

If the interface is between a fluid and vacuum, then too there is no boundary condition related to displacement as the fluid can move into the vacuum. The boundary condition related to stress will be only related to normal stress as fluid cannot sustain shear stress in any case. Hence,

$$\tau_{33/x_3=0} = 0. \tag{98b}$$

In the next section, we will show the derivation of reflection and transmission coefficients for an incident *SH* and *P* wave on a solid–solid boundary. We will assume that the waves are harmonic in nature. For the *SH* wave, the derivations can start with displacement components as there is no mode conversion. However, for the *P* wave, the derivations need to be started from displacement potentials as there will be mode conversions.

9.3. *Reflection and Transmission Coefficients for Solid–Solid Boundary for SH Wave*

Let there be a harmonic *SH* wave traveling in x_1x_3 plane and it falls on the boundary at an angle of i_1 (Figure 15). Hence, the angle of

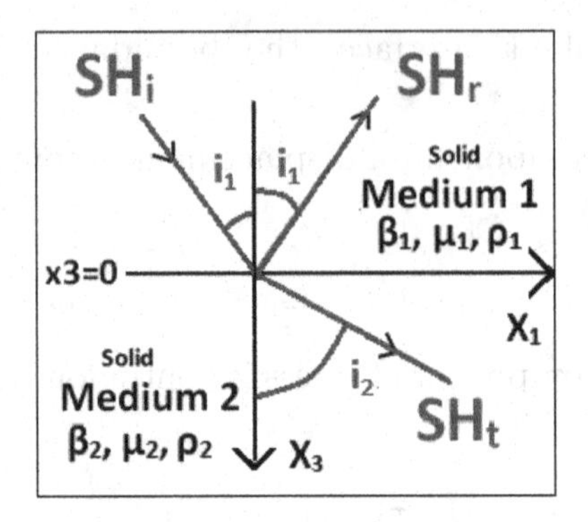

Figure 15. Interaction of an SH wave with an interface.

reflection is i_1. Let the angle of refraction/transmission be i_2. Here, the particle motion u will be in x_2 direction. As SH wave particle motion is orthogonal to that of P and SV waves, there will be no mode conversion. Hence, there will only be reflected and transmitted SH wave. Let the incident, reflected, and transmitted SH wave displacement for circular frequency ω and time t be u_{2i}, u_{2r}, and u_{2t}, respectively, with maximum amplitude of displacement as B_1, B_2, and B_3. We may write u_{2i}, u_{2r}, u_{2t} as

$$u_{2i} = B_1 e^{i\omega(px_1 + \eta_{\beta_1} x_3 - t)}, \tag{99a}$$

$$u_{2r} = B_2 e^{i\omega(px_1 - \eta_{\beta_1} x_3 - t)}, \tag{99b}$$

$$u_{2t} = B_3 e^{i\omega(px_1 + \eta_{\beta_1} x_3 - t)}, \tag{99c}$$

respectively, where β_1, μ_1, and ρ_1 are the S wave velocity, shear modulus, and density for medium 1 and β_2, μ_2 and ρ_2 are the corresponding parameters for medium 2. Here, ray parameter p and vertical slowness η_{β_1} and η_{β_2} for media 1 and 2, respectively, are given as

$$p = \frac{\sin i_1}{\beta_1} = \frac{\sin i_2}{\beta_2}, \tag{100a}$$

$$\eta_{\beta_1} = \frac{\cos i_1}{\beta_1}, \tag{100b}$$

$$\text{and} \quad \eta_{\beta_2} = \frac{\cos i_2}{\beta_2}. \tag{100c}$$

In this case, to estimate reflection and transmission coefficients, we need to apply only two conditions, *viz.* continuity of displacement component u_2 and shear stress τ_{23} across the boundary. The net displacement in medium 1 at time t will be

$$u_2^{(1)} = B_1 e^{i\omega(px_1 + \eta_{\beta_1} x_3 - t)} + B_2 e^{i\omega(px_1 - \eta_{\beta_1} x_3 - t)} \tag{101a}$$

and that in medium 2 will be

$$u_2^{(2)} = B_3 e^{i\omega(px_1 + \eta_{\beta_1} x_3 - t)}. \tag{101b}$$

The shear stress τ_{23} can be written as

$$\tau_{23} = 2\mu e_{23} = \mu \left[\frac{\partial u_2}{\partial x_3} + \frac{\partial u_3}{\partial x_2}\right] = \mu \left[\frac{\partial u_2}{\partial x_3}\right]. \tag{101c}$$

Applying Eq. (101(c)) on Eqs. (101(a)) and (101(b)), we get the shear stresses in medium 1 and 2 as follows:

$$\begin{aligned}\tau_{23}^{(1)} &= i\omega\mu_1\eta_{\beta_1}[B_1 e^{i\omega(px_1 + \eta_{\beta_1} x_3 - t)} \\ &\quad - B_2 e^{i\omega(px_1 - \eta_{\beta_1} x_3 - t)}],\end{aligned} \tag{101d}$$

$$\tau_{23}^{(2)} = i\omega\mu_2\eta_{\beta_2} B_3 e^{i\omega(px_1 + \eta_{\beta_1} x_3 - t)}. \tag{101e}$$

Using the boundary condition (i) $u_2^{(1)} = u_2^{(2)}$ at $x_3 = 0$ on Eqs. (101(a)) and (101(b)), we get

$$B_1 + B_2 = B_3. \tag{102a}$$

Using the boundary condition (ii) $\tau_{23}^{(1)} = \tau_{23}^{(2)}$ at $x_3 = 0$ on Eqs. (101(d)) and (101(e)) we get

$$(\mu_1\eta_{\beta_1} + \mu_2\eta_{\beta_2})B_2 = (\mu_1\eta_{\beta_1} - \mu_2\eta_{\beta_2})B_1. \tag{102b}$$

From Eqs. (102(a)) and (102(b)), we get the reflection coefficient R and transmission coefficient T as

$$R = \frac{B_2}{B_1} = \frac{(\mu_1\eta_{\beta_1} - \mu_2\eta_{\beta_2})}{(\mu_1\eta_{\beta_1} + \mu_2\eta_{\beta_2})} = \frac{\rho_1\beta_1\cos i_1 - \rho_2\beta_2\cos i_2}{\rho_1\beta_1\cos i_1 + \rho_2\beta_2\cos i_2} \tag{103a}$$

and

$$T = \frac{B_3}{B_1} = \frac{(2\mu_1\eta_{\beta_1})}{(\mu_1\eta_{\beta_1} + \mu_2\eta_{\beta_2})} = \frac{2\rho_1\beta_1\cos i_1}{\rho_1\beta_1\cos i_1 + \rho_2\beta_2\cos i_2} \tag{103b}$$

respectively. From Eqs. (103(a)) and (103(b)), we see that the reflection and transmission coefficients, i.e., the normalized amplitudes of reflected and transmitted waves depend on the medium properties of the two media on the two sides of the interface and on the angle of incidence.

9.4. *Reflection and Transmission Coefficients for Solid–Solid Boundary for Incident P Wave*

In this case, as there will be mode conversion, we start with displacement potential. From the Helmholtz's theorem, the displacement vector $\bar{u}$ can be represented as the sum of the gradient of scalar potential Φ and the curl of the vector potential Ψ (Eq. (19)). We have earlier seen that Φ represents the P wave motion (Eq. (21(b))) and Ψ represents the S wave motion (Eq. (21(c))). Let a P wave be incident on the boundary (Figure 16). This will give rise to reflected P, reflected S (converted) waves, transmitted P, and transmitted S (converted) waves. Let the angle of incidence and reflection for the P wave be i_1, that for the reflected S wave be j_1, angle of refraction/transmission for P and converted S waves be i_2 and j_2, respectively. Let the P and S wave velocity, shear modulus and density of medium 1 be α_1, β_1, μ_1 and ρ_1 respectively and that for medium 2 be α_2, β_2, μ_2 and ρ_2, respectively (Figure 16). Let Φ_i, Φ_r, and Φ_t represent displacement potentials for the incident, reflected, and transmitted P waves, and φ_r and φ_t represent displacement potentials for the reflected and

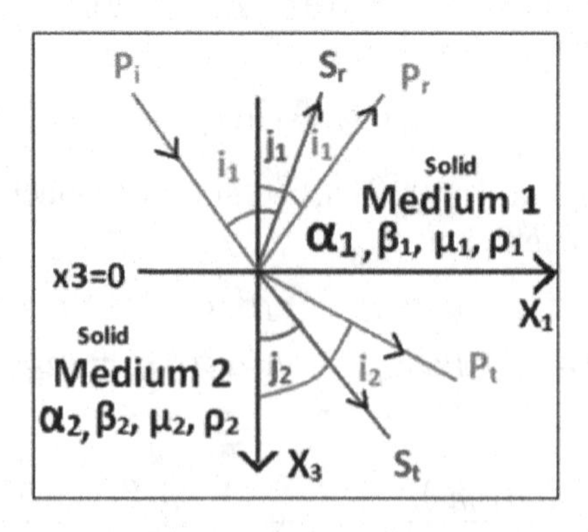

Figure 16. Interaction of a P wave with an interface.

transmitted S waves, respectively. Let A_1, A_2, A_3, B_2, and B_3 represent maximum amplitudes of the incident P, reflected P, transmitted P, reflected S, and transmitted S, respectively. Let p, η_{α_1}, η_{α_2}, η_{β_1}, and η_{β_2} be ray parameter, vertical slowness for media 1 and 2 for P and vertical slowness for media 1 and 2 for S wave, respectively. Here, the raypaths as well as particle motion of P as well as S (here SV) are all confined in the x_1x_3 plane. Hence, only the displacement components u_1 and u_3 are to be considered. Assuming all waves are harmonic, we may write

$$\emptyset_i = A_1 e^{i\omega(px_1+\eta_{\alpha_1}x_3-t)}, \tag{104a}$$

$$\emptyset_r = A_2 e^{i\omega(px_1-\eta_{\alpha_1}x_3-t)}, \tag{104b}$$

$$\emptyset_t = A_3 e^{i\omega(px_1+\eta_{\alpha_2}x_3-t)}, \tag{104c}$$

$$\varphi_r = B_2 e^{i\omega(px_1-\eta_{\beta_1}x_3-t)}, \tag{104d}$$

$$\varphi_t = B_3 e^{i\omega(px_1+\eta_{\beta_1}x_3-t)}. \tag{104e}$$

The requisite boundary conditions related to displacements are

$$u_3^{(1)}|_{x3=0} = u_3^{(2)}|_{x3=0}, \tag{105a}$$

$$u_1^{(1)}|_{x3=0} = u_1^{(2)}|_{x3=0}, \tag{105b}$$

and those related to stress components are

$$\tau_{33}^{(1)}|_{x3=0} = \tau_{33}^{(2)}|_{x3=0}, \tag{105c}$$

$$\tau_{13}^{(1)}|_{x3=0} = \tau_{13}^{(2)}|_{x3=0}. \tag{105d}$$

Now, the scalar and vector displacement potentials in the two media are

$$\emptyset_1 = A_1 e^{i\omega(px_1+\eta_{\alpha_1}x_3-t)} + A_2 e^{i\omega(px_1-\eta_{\alpha_1}x_3-t)}, \tag{106a}$$

$$\emptyset_2 = A_3 e^{i\omega(px_1+\eta_{\alpha_2}x_3-t)}, \tag{106b}$$

$$\varphi_1 = B_2 e^{i\omega(px_1-\eta_{\beta_1}x_3-t)}, \tag{106c}$$

$$\varphi_2 = B_3 e^{i\omega(px_1+\eta_{\beta_2}x_3-t)}. \tag{106d}$$

Using Helmholtz's equation (Eq. (19)), the u_1 displacement component in the two media can be written as

$$u_1^{(1)} = \frac{\partial \emptyset_1}{\partial x_1} - \frac{\partial \varphi_1}{\partial x_3}, \tag{107a}$$

$$u_1^{(2)} = \frac{\partial \emptyset_2}{\partial x_1} - \frac{\partial \varphi_2}{\partial x_3}. \tag{107b}$$

Using the boundary condition given in Eq. (105(b)) on Eqs. (107(a)) and (107(b)), we get

$$p(A_1 + A_2) + \eta_{\beta_1} B_2 = pA_3 - \eta_{\beta_2} B_3. \tag{108}$$

Again, using Eq. (19), the u_3 displacement component in the two media can be written as

$$u_3^{(1)} = \frac{\partial \emptyset_1}{\partial x_3} + \frac{\partial \varphi_1}{\partial x_1}, \tag{109a}$$

$$u_3^{(2)} = \frac{\partial \emptyset_2}{\partial x_3} + \frac{\partial \varphi_2}{\partial x_1}. \tag{109b}$$

Using the boundary condition given in Eq. (105(a)) on Eqs. (109(a)) and (109(b)), we get

$$\eta_{\alpha_1}(A_1 - A_2) + pB_2 = \eta_{\alpha_2} A_3 + pB_3. \tag{110a}$$

Applying $\tau_{13}^{(1)} = \tau_{13}^{(2)}$ (boundary condition given in Eq. (105(d))), we get

$$\begin{aligned} &\mu_1[2p\eta_{\alpha_1}(A_1 - A_2) + (p^2 - \eta_{\beta_1}^2)B_2] \\ &\quad = \mu_2[2p\eta_{\alpha_2}A_3 + (p^2 - \eta_{\beta_2}^2)B_3] \end{aligned} \tag{110b}$$

and applying $\tau_{33}^{(1)} = \tau_{33}^{(2)}$ (boundary condition given in Eq. (105(c))), we get

$$\begin{aligned} &\rho_1[(2\beta_1^2 p^2 - 1)(A_1 + A_2) + (2\beta_1^2 p\eta_{\beta 1})B_2] \\ &\quad = \rho_2[(2\beta_2^2 p^2 - 1)A_3 - 2\beta_2^2 p\eta_{\beta 2}B_3]. \end{aligned} \tag{110c}$$

Dividing both the left-hand and right-hand sides of Eqs. (108), (110(a)), (110(b)), and (110(c)), and writing the four equations in matrix format, we get the following matrix (Eq. (111)), where $A_2/A_1 = Rpp$ is the reflection coefficient when the incident P wave is reflected as P wave, $A_3/A_1 = Tpp$ is the transmission coefficient when the incident P wave is transmitted as P wave, $B_2/A_1 = Rps$ is the reflection coefficient when the incident P wave is reflected as S wave, and $B_3/A_1 = Tps$ is the transmission coefficient when the incident P wave is transmitted as S wave. It is observed that all these parameters are dependent on the properties of the two media and the angle of incidence. By solving Eq. (111), we can estimate the reflection and transmission coefficients when a P wave is incident upon a solid–solid boundary. In the same fashion, for any type of interface and incident wave using appropriate boundary condition relationships for reflection and transmission coefficients may be estimated. These equations are routinely used for estimation of medium properties of the Earth using records of reflected and/or transmitted wave amplitudes:

$$\begin{bmatrix} 1 & -1 & \dfrac{\eta_{\beta_1}}{p} & \dfrac{\eta_{\beta_2}}{p} \\ 1 & -\dfrac{\eta_{\alpha_2}}{\eta_{\alpha_1}} & -\dfrac{p}{\eta_{\alpha_1}} & \dfrac{p}{\eta_{\alpha_1}} \\ 1 & -\dfrac{\mu_2\eta_{\alpha_2}}{\mu_1\eta_{\alpha_1}} & -\dfrac{p^2-\eta_{\beta_1}^2}{p\eta_{\alpha_1}} & \dfrac{\mu_2(p^2-\eta_{\beta_1}^2)}{\mu_1 p\eta_{\alpha_1}} \\ 1 & -\dfrac{\rho_2(2\beta_2^2p^2-1)}{\rho_1(2\beta_1^2p^2-1)} & \dfrac{2\beta_1^2p^2\eta_{\beta_1}}{(2\beta_1^2p^2-1)} & \dfrac{2\rho_2\beta_2^2p\eta_{\beta_2}}{\rho_1(2\beta_1^2p^2-1)} \end{bmatrix} \begin{bmatrix} \dfrac{A_2}{A_1} \\ \dfrac{A_3}{A_1} \\ \dfrac{B_2}{A_1} \\ \dfrac{B_3}{A_1} \end{bmatrix} = \begin{bmatrix} 1 \\ 1 \\ 1 \\ -1 \end{bmatrix}. \tag{111}$$

References

[1] Sokolnikoff, I. S. (1956). *Mathematical Theory of Elasticity*, McGraw Hill: New York, p. 476.

[2] Lay, T. and Wallace, T. C. (1995). *Modern Global Seismology*, Academic Press: Cambridge, MA, p. 521.

[3] Aki, K. and Richards, P. G. (2002). *Quantitative Seismology*, University Science Books: Sausalito, CA, p. 700.

[4] Bath, M. (1968). *Mathematical Aspects of Seismology*, Elsevier Publishing Co.: Amsterdam, Netherlands, p. 415.

[5] Bullen, K. E. and Bolt, B. A. (1985). *An Introduction to the Theory of Seismology*, Cambridge University Press: Cambridge, p. 499.

[6] Shearer, P. M. (2009). *Introduction to Seismology*, Cambridge University Press: Cambridge, p. 396.
[7] Stein, S. and Wysession, M. (2003). *An Introduction to Seismology, Earthquakes, and Earth Structure*, Blackwell Publishing: Oxford, p. 498.
[8] Udias, A. (1999). *Principles of Seismology*, Cambridge University Press: Cambridge, p. 475.

https://doi.org/10.1142/9789811245367_0006

Chapter 6

Mathematical Study of Reflection and Transmission Phenomenon of Plane Waves at the Interface of Two Dissimilar Initially Stressed Rotating Micro-Mechanically Modeled Piezoelectric Fiber-Reinforced Composite Half-spaces

Abhishek Kumar Singh* and Sayantan Guha†

Department of Mathematics and Computing,
Indian Institute of Technology (Indian School of Mines),
Dhanbad, Jharkhand, India
**abhi.5700@gmail.com*
†sayantanguha.maths@gmail.com

Abstract

This chapter has two objectives: (1) to present the micro-mechanics model of Piezoelectric Fiber-Reinforced Composites (PFRCs) and illustrate some of its advantages and (2) to analytically study the impacts of normal/shear initial stresses and rotation on energies carried by different reflected/transmitted waves at the interface of two dissimilar PFRCs. Numerical studies are performed on PFRCs comprised of PZT-5A-epoxy combination and CdSe-epoxy combination, which are modeled employing Strength of Materials (SM) technique with Rule of Mixtures (RM). Some electro-mechanical advantages of PFRC over monolithic piezoelectric materials are demonstrated. Due to incidence of a quasi-longitudinal (qP) wave, three reflected/transmitted waves, *viz.* quasi-longitudinal (qP), quasi-transverse (qSV), and electro-acoustic (EA) waves are generated in the PFRCs. The propagation directions of all reflected/transmitted waves are graphically demonstrated. The closed-form expressions of amplitude ratios of all reflected/transmitted waves are derived utilizing appropriate electro-mechanical boundary conditions. As the amplitude ratios cannot be used exclusively to validate the numerical results, the expressions of energy ratios of all reflected/transmitted waves and interaction energy are derived, which exhibit the influence of existing parameters, and the law of conservation of energy is established. This work presents a novel effort to develop a connection between deriving the PFRC's micro-mechanical model and analyzing the wave reflection/transmission phenomenon in it.

Keywords: Piezoelectric Fiber-Reinforced Composite (PFRC), reflection, transmission, rotation, initial stress, amplitude ratio, energy ratio

1. Introduction

Investigation of the phenomenon of wave reflection/transmission has always remained a subject of prime importance owing to its necessity in areas such as earthquake engineering, seismic exploration, and geophysics, to name a few. Proper analyzes of wave reflection/transmission are crucial for a better understanding of the composition of the Earth's internal structure as waves propagating through a medium carry a significant amount of information about that medium. However, they are not limited to the geophysical aspects as the analyzed results are also vital for tremendous applications in fields such as mining and acoustics, among others. Several relevant works on wave reflection/transmission phenomenon in anisotropic materials and composite-layered structures can be found in the existing literature [1–3].

At the hands of Curie and Curie [4], the piezoelectricity phenomenon came to be known. Piezoelectricity exists in non-centrosymmetric crystals, enabling them to generate and accumulate electric charge on the application of external mechanical stress (direct effect) and causes mechanical strain in an applied electrical field (inverse effect). This property is extensively used to control the electrical and mechanical outputs of various structures using smart materials. The direct effect finds its uses in applications such as pressure sensors, rotation sensors, microphones, ultrasonic detectors, noise and vibration control, and SONAR. The inverse effect is utilized in both low-frequency applications, such as electronic buzzers, actuators, and speakers, as well as high-frequency applications, such as pumps, motors, and drills. Certain applications such as transformers, quartz crystal oscillators, ultrasonic nondestructive testing, balance, and AFM probe employ both the direct and inverse effects. Several noteworthy studies regarding wave reflection/transmission phenomenon and wave propagation in smart composites were conducted by Yuan and Zhu [5], Pang *et al.* [6], and Guha *et al.* [7], among many.

The incorporation of initial stresses in studies of wave reflection/transmission phenomenon and wave propagation is essential due to their impacts on the different materials' responses considered. The occurrence of initial stress is often inevitable due to many reasons, such as quenching process, non-uniform material properties, machining, shrinkage and/or growth during processing and cooling down to operating/room temperature, enhancing fracture toughness, slow creep deformation, gravity, and thermal/chemical influences. Also, to prevent brittle fracture, pre-stress is imposed on layered piezoelectric structures during manufacturing processes, which may lead to degradation, microcracking, debonding, and delamination of the considered layer, among other things. Also, keeping in mind practical engineering applications, it is essential to analyze the nature of waves when they propagate through a rotating media. It is known that in a rotating elastic or piezoelectric structure, the Coriolis and centrifugal forces change the speed of wave or vibration frequency. This phenomenon is highly useful in designing and using rotation/angular rate sensors (gyroscopes). The piezoelectric gyroscopes of rotating motion sensors have extensive applications in motion cameras, navigation, automobiles, robotics, machine control, etc. They can also be used for rotation-induced frequency shifts in Bulk Acoustic Wave (BAW) or Surface Acoustic Wave (SAW) devices for measuring angular rates. Thus, to enhance the gyroscope's working capability, the dynamics of a piezoelectric body in a rotating frame needs to be analyzed with precision considering the effects of Coriolis and centrifugal forces. Hence, it becomes quintessential to study the influences of rotation and initial stresses meticulously, as has been conducted by several eminent researchers [8–11]. However, it is found from most of the previous work in the extant literature that only influences of normal initial stresses were taken into consideration, whereas in reality, initial stresses affect the considered structures from all sides and not just along the normal direction. Thus, the authors have examined the impacts of normal initial stresses $(\tau_{11}^{0}, \tau_{33}^{0})$ and shear initial stress (τ_{13}^{0}), keeping practical aspects in mind.

With the emergence of more superior technological advancements, development of composite structures is ongoing at a massive pace due to their uses in major scientific and engineering fields. More specifically, among several types of composites, Piezoelectric Fiber-Reinforced Composites (PFRCs) are of enormous significance at present times owing to certain significant advantages, such as strength and lightweight. Thus, PFRCs offer superior performance compared to monolithic piezoelectric materials as they can be optimized to enhance the properties desired in respective applications. Consequently, continuous attempts are being made to develop and improvise such materials focusing on their effective commercial utilization in countless areas, including aeronautics, sports, constructions (bridges, buildings, etc.), remote explorations, medical services (pressure and heartbeat monitors), accelerometers, and ultrasonic imaging. The prediction of the electro-mechanical properties of PFRCs has been a dynamic research area for several years. Consequently, some notable works on developing the micro-mechanics of composite materials by analytical and numerical methods making use of techniques, such as strength of materials, asymptotic homogenization method, continuum mechanics, method of cells, and finite-element method, have been performed in the past [12–15]. The effective mechanical and electrical traits obtained from the different studies were found to resemble each other closely — this validated the physical and mathematical assumptions pertaining to the studies.

Due to the apparently endless advantages of PFRC in comparison to monolithic piezoelectric materials, several relevant works dealing with wave propagation phenomenon in fiber-reinforced composites are found in the literature [16, 17] and some works on wave reflection phenomenon in fiber-reinforced composites are also conducted [18, 19]. Literature survey reveals that no mathematical studies have been performed yet on the wave reflection/transmission phenomenon in a composite structure comprised of two dissimilar PFRCs with complexities such as rotation and normal/shear initial stresses. The present chapter is framed to explore the same for contemplating the phenomenon in constructed smart structures fully. Therefore, this study presents a novel effort for shedding some light on the yet uninvestigated areas of the field and develop a connection between derivation of the composite's micro-mechanics model and analysis of wave reflection/transmission phenomenon in it.

2. The Micro-Mechanics Modeling of PFRC

Both of the considered PFRC media in this work are comprised of piezoelectric fibers encompassed by a matrix material which is piezoelectrically inactive. For the composition of such a structure, certain assumptions are taken into account, which are encapsulated as follows:

- The composite structure is homogeneous.
- The fibers and matrix are in rigid contact with each other, and no slippage occurs among them.
- The fibers positioned along the x-axis are continuous and parallel.
- The changes in the electric field are identical in both matrix and fibers.
- The composite is in the presence of a constant electric field acting along both transverse and longitudinal directions to the fiber.

In the micro-mechanics approach, the complete PFRC may be thought to be an accumulation of Rectangular Representative Volume (RVE) elements consisting of parts of the fiber and the adjoining matrix material as in Ref. [14]. The effective material properties of the PFRC are likely to be equivalent to those of the RVE in an averaged sense, and the properties can be studied meticulously by analyzing the RVE. The concept of RVE can be easily observed and realistically envisioned from Figure 3(c). The mathematical analysis carried out in this work regarding the derivation of the expressions of the PFRC material constants (using SM and RM techniques) is valid, in general, and has no dependence on the fiber's cross-section shape, which may be circular, rectangular, etc. The constitutive equations for piezoelectric fibers are given in compact form as follows:

$$\{\tau^f\} = [c^f]\{s^f\} - [e^f]^T\{E^f\}, \quad \{D^f\} = [e^f]\{s^f\} + [\varepsilon^f]\{E^f\}, \quad (1)$$

and the same for the piezoelectrically inactive matrix are

$$\{\tau^m\} = [c^m]\{s^m\}, \quad \{D^m\} = [\varepsilon^m]\{E^m\}, \tag{2}$$

where

$$\{\tau^\gamma\} = [\tau_{11}^\gamma \; \tau_{22}^\gamma \; \tau_{33}^\gamma \; \tau_{23}^\gamma \; \tau_{13}^\gamma \; \tau_{12}^\gamma]^T,$$

$$\{s^\gamma\} = [s_{11}^\gamma \; s_{22}^\gamma \; s_{33}^\gamma \; 2s_{23}^\gamma \; 2s_{13}^\gamma \; 2s_{12}^\gamma]^T,$$

$$[c^\gamma] = \begin{bmatrix} c_{11}^\gamma & c_{12}^\gamma & c_{13}^\gamma & 0 & 0 & 0 \\ c_{12}^\gamma & c_{22}^\gamma & c_{23}^\gamma & 0 & 0 & 0 \\ c_{13}^\gamma & c_{23}^\gamma & c_{33}^\gamma & 0 & 0 & 0 \\ 0 & 0 & 0 & c_{44}^\gamma & 0 & 0 \\ 0 & 0 & 0 & 0 & c_{55}^\gamma & 0 \\ 0 & 0 & 0 & 0 & 0 & c_{66}^\gamma \end{bmatrix},$$

$$[e^f] = \begin{bmatrix} 0 & 0 & e_{31}^f \\ 0 & 0 & e_{32}^f \\ 0 & 0 & e_{33}^f \\ 0 & e_{24}^f & 0 \\ e_{15}^f & 0 & 0 \\ 0 & 0 & 0 \end{bmatrix}^T, \quad \{D^\gamma\} = \begin{bmatrix} D_1^\gamma \\ D_2^\gamma \\ D_3^\gamma \end{bmatrix},$$

$$\{E^\gamma\} = \begin{bmatrix} E_1^\gamma \\ E_2^\gamma \\ E_3^\gamma \end{bmatrix}, \quad [\varepsilon^y] = \begin{bmatrix} \varepsilon_{11}^y & 0 & 0 \\ 0 & \varepsilon_{22}^y & 0 \\ 0 & 0 & \varepsilon_{33}^y \end{bmatrix}.$$

In the above expressions, the superscript T represents the transpose of the matrix, and the superscript γ is equal to f and m denoting the components of the fiber part and the matrix part, respectively. The terms in Eqs. (1) and (2) and some of the terms used in subsequent sections are explained in Table 1.

Since the electric field is equal in both the matrix and the fiber, we get $E_i^m = E_i^f = E$. As the fiber and matrix are bonded to each other perfectly, resultant strains are equal in the x-direction, which implies $s_{11}^f = s_{11}^m = s_{11}$. On employing RM, the different expressions

Table 1. Nomenclature.

Term	Representation	Term	Representation
τ_{ij}	Stress tensor	c_{ij}	Elastic constants
s_{ij}	Strain tensor	e_{ij}	Piezoelectric parameters
E_i	Electric field	D_i	Electric displacement
ε_{ij}	Electric permittivity	Ω	Angular rotation rate
τ_{ij}^0	Initial stresses	ρ	Density
ε_{ijk}	Permutation symbol		

of strains are obtained, following [14], as

$$\begin{aligned} &s_{22} = \Omega_f s_{22}^f + \Omega_m s_{22}^m, \quad s_{33} = \Omega_f s_{33}^f + \Omega_m s_{33}^m, \\ &s_{23} = \Omega_f s_{23}^f + \Omega_m s_{23}^m, \quad s_{13} = \Omega_f s_{13}^f + \Omega_m s_{13}^m, \\ &s_{12} = \Omega_f s_{12}^f + \Omega_m s_{12}^m. \end{aligned} \tag{3}$$

Here Ω_f and Ω_m are used to symbolize the fiber volume fraction and the matrix volume fraction, respectively, which are related by $\Omega_f + \Omega_m = 1$. Thus, $\Omega_f = 1$ means that only piezoelectric fibers are present and $\Omega_m = 1$ means that only epoxy is present.

The average composite stress along the x-axis is conveyed with respect to the average stresses in matrix and fiber using RM for the PFRC to achieve axial equilibrium. Again, for equilibrium consideration, as in [20], the average stresses acting laterally along the y- and z-axes and the average shear stresses are equal in matrix and fiber and are identical to the composite's average stresses. Thus, the stresses can be expressed as

$$\begin{aligned} &\tau_{11} = \Omega_f \tau_{11}^f + \Omega_m \tau_{11}^m, \quad \tau_{22} = \tau_{22}^f = \tau_{22}^m, \quad \tau_{33} = \tau_{33}^f = \tau_{33}^m, \\ &\tau_{23} = \tau_{23}^f = \tau_{23}^m, \quad \tau_{13} = \tau_{13}^f = \tau_{13}^m, \quad \tau_{12} = \tau_{12}^f = \tau_{12}^m. \end{aligned} \tag{4}$$

On employing RM, the resultant electric displacements in the PFRC are

$$\begin{aligned} &D_1 = \Omega_f D_1^f + \Omega_m D_1^m, \quad D_2 = \Omega_f D_2^f + \Omega_m D_2^m, \\ &D_3 = \Omega_f D_3^f + \Omega_m D_3^m, \end{aligned} \tag{5}$$

and the density ρ of the PFRC is obtained, following [21], as

$$\rho = \Omega_f \rho^f + \Omega_m \rho^m. \tag{6}$$

The expressions of all the constants occurring in the PFRC constitutive equations are obtained using the SM technique and RM approach and are presented in Appendix. It can be readily observed that the different electro-mechanical data of the PFRCs are solely dependent on Ω_f and Ω_m. Hence, on changing the values of Ω_f and Ω_m, different sets of data for the PFRCs can be derived, and the most advantageous electro-mechanical properties can be utilized for some particular values of Ω_f and Ω_m, as required in scientific and engineering applications. In this chapter, the two distinct PFRC media considered for half-spaces H_1 and H_2 are comprised of PZT-5A fiber-epoxy matrix combination and CdSe fiber-epoxy matrix combination, respectively. Variations of some of the electro-mechanical properties of both PFRCs are graphically demonstrated in Figures 1 and 2.

Here, $\overline{e_{31}} = e_{31}/e_{31}^f$, $\overline{\epsilon_{ii}} = \epsilon_{ii}/\epsilon_{ii}^f$, $(i = 1, 3)$, $\overline{c_{ij}} = c_{ij}/c_{ij}^f$, and $\overline{\rho} = \rho/\rho^f$ are used for scaling purposes. It is observed from Figures 1 and 2 that the PFRCs generate maximum electrical response in comparison to their monolithic piezoelectric parts for $\Omega_f = 0.9$ and $\Omega_m = 0.1$, respectively. These electrical properties are of enormous importance as they govern the performances of various piezoelectric devices like actuators and sensors. The dielectric permittivities are vital for influencing voltage developed in piezoelectric sensors and

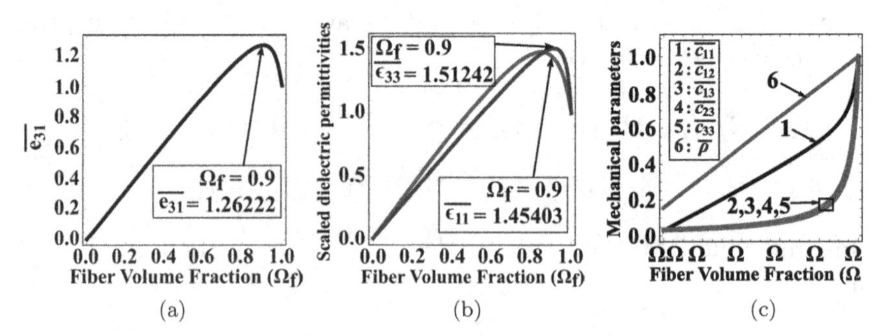

(a) (b) (c)

Figure 1. Scaled (a) piezoelectric constant $\overline{e_{31}}$, (b) dielectric permittivities $(\overline{\epsilon_{11}}, \overline{\epsilon_{33}})$, and (c) stiffness constants $(\overline{c_{ij}})$ and density $(\overline{\rho})$ for PZT-5A-epoxy combination.

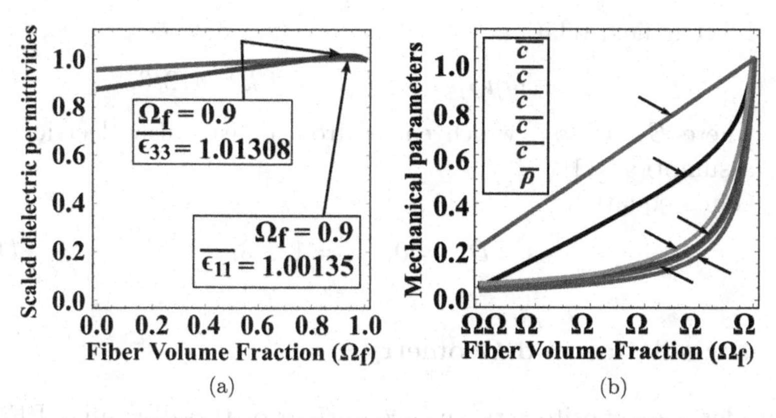

(a) (b)

Figure 2. Scaled (a) dielectric permittivities ($\overline{\epsilon_{11}}, \overline{\epsilon_{33}}$) and (b) stiffness constants ($\overline{c_{ij}}$) and density ($\overline{\rho}$) for CdSe-epoxy combination.

capacitor applications. It is seen that the magnitudes of densities and stiffness parameters for the PFRCs lie between the values of the matrix part and their corresponding PE fiber part. This results in low density and mechanically flexible structures that are essential for designing electro-elastic transducers.

3. Basic Constitutive Equations

The constitutive equations for the PFRC in the absence of body forces and electric forces are as follows.

(A) Strain–displacement relations:

$$s_{ij} = (u_{i,j} + u_{j,i})/2, \;\; i, j = 1, 2, 3. \tag{7}$$

(B) Stress–strain–electric field relations:

$$\tau_{ij} = c_{ijkl}s_{kl} - e_{kij}E_k, \quad i, j, k, l = 1, 2, 3. \tag{8}$$

(C) Equation of motion:

$$\tau_{ji,j} + (u_{i,k}\tau^0_{kj})_{,j} = \rho[\ddot{u}_i + 2\varepsilon_{ijk}\Omega_j\dot{u}_k + \Omega_i\Omega_j u_j - \Omega_j^2 u_i], \quad i, j, k = 1, 2, 3. \tag{9}$$

(D) Electric field relation:

$$D_i = e_{ijk}s_{jk} + \varepsilon_{ij}E_j, \quad i,j,k = 1,2,3, \tag{10}$$

where $E_i = -\phi_{,i}$, which comes from quasi-static electric field assumption [11].

(E) Gauss equation:

$$D_{i,i} = 0, \quad i = 1,2,3. \tag{11}$$

4. Formulation and Geometry

Consider a composite structure comprised of two dissimilar PFRC half-spaces. The overall structure, which is poled along the z-direction, remains undisturbed. According to the considered Cartesian coordinate system, the x-axis is lying along the interface, while the z-axis is pointing vertically downwards. The point "O" demarcates the plane wave's point of incidence (Figure 3). It is considered that the structure is rotating about the z-axis with constant angular velocity Ω and is under the influence of normal and shear initial stress $(\tau_{11}^0, \tau_{33}^0, \tau_{13}^0)$ with different directions of compaction and rarefaction.

We analyze our problem considering the xz-plane and assume a plane strain condition owing to which we have $\partial(\cdot)/\partial y = 0$, i.e., all considered functions will depend solely on x and z directions and time t. Consider that $\{u^{(1)} = u^{(1)}(x,z,t),\ v^{(1)} = 0,\ w^{(1)} = w^{(1)}(x,z,t)\}$ and $\phi^{(1)}(x,z,t)$ indicate the mechanical displacements and electric potential in the lower half-space (H_1), respectively, where wave incidence and reflections occur. Similarly, let $\{u^{(2)} = u^{(2)}(x,z,t),\ v^{(2)} = 0,\ w^{(2)} = w^{(2)}(x,z,t)\}$ and $\phi^{(2)}(x,z,t)$ denote the components of the mechanical displacements and electric potential in the upper half-space (H_2), respectively, where wave transmissions occur. The governing equations neglecting body forces and electric forces are given by

$$\begin{aligned}(c_{11} + \tau_{11}^0)u_{,11} &+ (c_{44} + \tau_{33}^0)u_{,33} + 2\tau_{13}^0 u_{,13} \\ &+ (c_{13} + c_{44})w_{,13} + (e_{31} + e_{15})\phi_{,13} = \rho(\ddot{u} - \Omega^2 u),\end{aligned} \tag{12}$$

$$\begin{aligned}(c_{13} + c_{44})u_{,13} &+ (c_{44} + \tau_{11}^0)w_{,11} + (c_{33} + \tau_{33}^0)w_{,33} \\ &+ 2\tau_{13}^0 w_{,13} + e_{15}\phi_{,11} + e_{33}\phi_{,33} = \rho\ddot{w},\end{aligned} \tag{13}$$

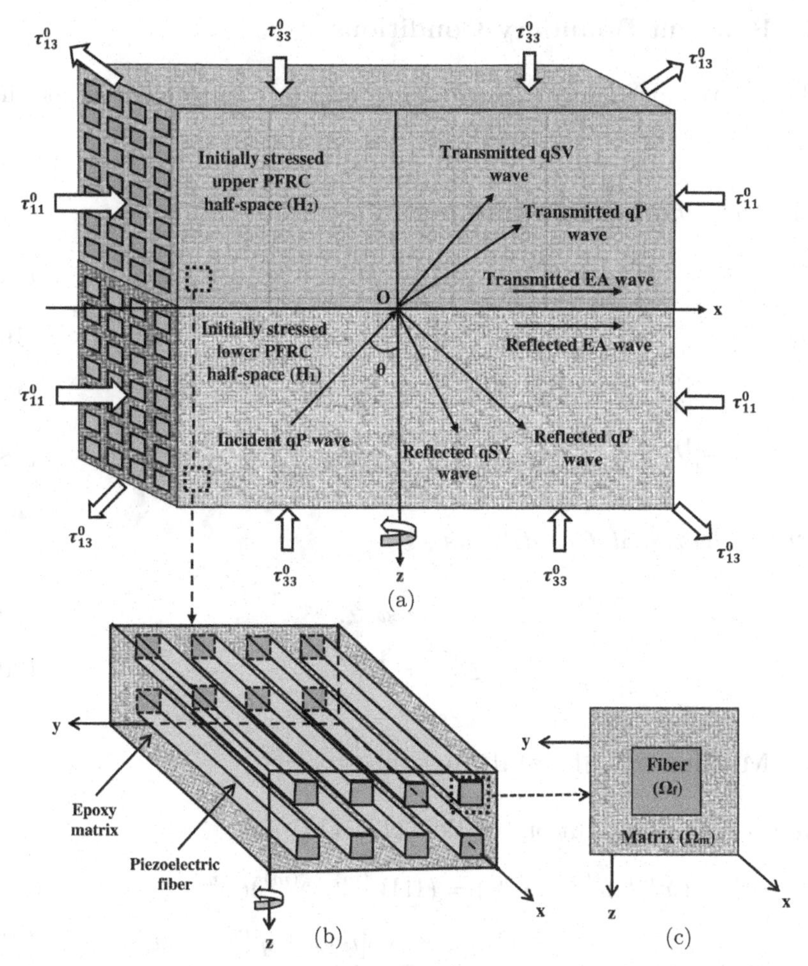

Figure 3. Geometry showing (a) the incident, reflected, and transmitted waves in the initially stressed dissimilar rotating PFRC half-spaces, (b) the micromechanics model of PFRC, and (c) the transverse cross-sectional area of a RVE of PFRC consisting of fiber and matrix volume parts.

$$
\begin{aligned}
&(e_{15}+e_{31})u_{,13}+e_{15}w_{,11}+e_{33}w_{,33}-\varepsilon_{11}\phi_{,11} \\
&\quad -\varepsilon_{33}\phi_{,33}=0.
\end{aligned}
\tag{14}
$$

The superposed dots used in Eqs. (12)–(14) represent differentiation with respect to time, and the comma notation denotes spatial derivatives.

5. Relevant Boundary Conditions

The electro-mechanical boundary conditions implemented at the interface (x-axis) are as follows.

5.1. *Mechanical Conditions*

$$u^{(1)} = u^{(2)}, \tag{15}$$

$$w^{(1)} = w^{(2)}, \tag{16}$$

$$\tau_{13}^{(1)} + \tau_{13}^{0} u_{,1}^{(1)} + \tau_{33}^{0} u_{,3}^{(1)} = \tau_{13}^{(2)} + \tau_{13}^{0} u_{,1}^{(2)} + \tau_{33}^{0} u_{,3}^{(2)}, \tag{17}$$

$$\tau_{33}^{(1)} + \tau_{13}^{0} w_{,1}^{(1)} + \tau_{33}^{0} w_{,3}^{(1)} = \tau_{33}^{(2)} + \tau_{13}^{0} w_{,1}^{(2)} + \tau_{33}^{0} w_{,3}^{(2)}. \tag{18}$$

5.2. *Electrical Conditions*

$$\phi^{(1)} = \phi^{(2)}, \tag{19}$$

$$D_3^{(1)} = D_3^{(2)}. \tag{20}$$

6. Mathematical Solution

We presume the solutions of Eqs. (12)–(14) as

$$(u^{(m)}, w^{(m)}, \phi^{(m)}) = (1, W^{(m)}, \Phi^{(m)}) U^{(m)} \times \exp[ik(x + q^{(m)} z - ct)], \tag{21}$$

where $k(= k_x)$ denotes the apparent wave number and c denotes the apparent phase velocity of the wave. The above solution form takes into consideration the generalized Snell's law, which leads to the fact that the apparent wave numbers of the different waves are equal. This particular form of the solution also ensures proper analysis of the natures of all reflected and transmitted waves, including EA waves whose velocity expressions cannot be derived due to quasi-static electric field assumption ($E_i = -\phi_{,i}$). The superscript m takes values 1 and 2, which represent half-spaces H_1 and H_2, respectively. Also, $W^{(m)}$ and $\Phi^{(m)}$ denote the ratios of mechanical displacement

$w^{(m)}$ and electric potential $\phi^{(m)}$ to the mechanical displacement $u^{(m)}$, respectively. Adhering to the generalized Snell's law, it is also supposed that $q^{(m)} = k_z/k_x$ denotes the projection ratio of the wave number on the z- and x-axes. We have $\mathrm{Re}(q^{(m)}) = \cot(\theta)$ if the wave propagates along the positive direction of the z-axis, making an angle θ. Making use of Eq. (21) in Eqs. (12)–(14), we acquire a coupled system of equations as follows:

$$
\begin{aligned}
&c_{11}^{(m)} + \tau_{11}^{0} + (c_{44}^{(m)} + \tau_{33}^{0})(q^{(m)})^2 + 2q^{(m)}\tau_{13}^{0} - \rho c^2(1+\Gamma^2)\\
&\quad + (c_{13}^{(m)} + c_{44}^{(m)})q^{(m)}W^{(m)} + (e_{31}^{(m)} + e_{15}^{(m)})q^{(m)}\Phi^{(m)} = 0,\\
&(c_{13}^{(m)} + c_{44}^{(m)})q^{(m)} + (c_{44}^{(m)} + \tau_{11}^{0} + (c_{33}^{(m)} + \tau_{33}^{0})(q^{(m)})^2\\
&\quad + 2q^{(m)}\tau_{13}^{0} - \rho c^2)W^{(m)} + (e_{15}^{(m)} + e_{33}^{(m)}(q^{(m)})^2)\Phi^{(m)} = 0,\\
&(e_{15}^{(m)} + e_{31}^{(m)})q^{(m)} + (e_{15}^{(m)} + e_{33}^{(m)}(q^{(m)})^2)W^{(m)}\\
&\quad - (\varepsilon_{11}^{(m)} + \varepsilon_{33}^{(m)}(q^{(m)})^2)\Phi^{(m)} = 0,
\end{aligned} \tag{22}
$$

In the aforementioned system of Eq. (22), the term Ω is transformed to $\Gamma(= \Omega/kc)$, which is also known as the inverse of Kibel number [22]. A non-trivial solution is obtained from the system in Eq. (22), provided the determinant of the coefficients of $1, W^{(m)}, \Phi^{(m)}$equated to zero. This results in a hexic polynomial in $q^{(m)}$, from which the expressions of the velocities of qP and qSV waves can be derived, which depend on θ. Fixing the expression of qP wave, the hexic polynomial equation in $q^{(m)}$ admits six solutions for $q^{(m)}$, which allow us to ascertain the directions of propagation of all reflected waves ($m = 1$) and transmitted waves ($m = 2$). The angles of the various reflected and transmitted waves, without the effects of initial stresses and rotation, are shown in Figure 4.

It can be observed from Figure 4(a) that the angles of the reflected qP wave and the incident qP wave are equal. The angle of the reflected qSV wave takes progressively lesser values as compared to qP wave. This is because qSV wave is slower than qP wave. The reflected EA wave, being a surface wave, propagates along the interface (x-axis).

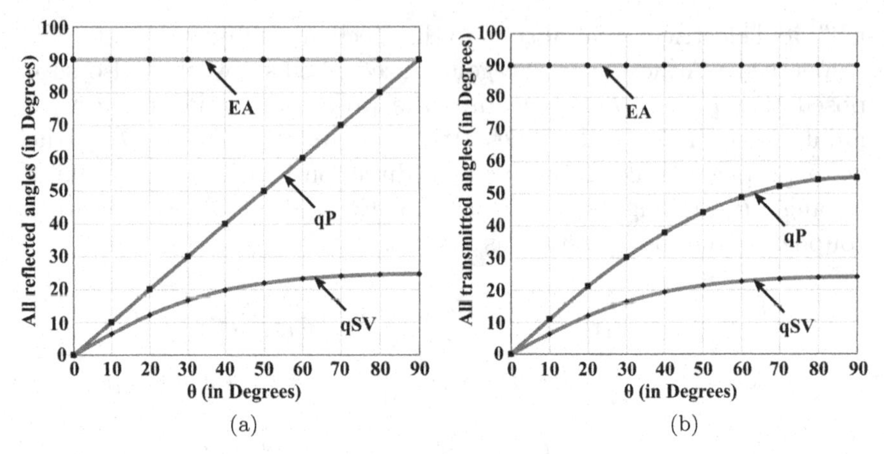

Figure 4. Variation of angles of all (a) reflected waves and (b) transmitted waves for incident qP wave.

From Figure 4(b), it is quite evident that the angles of the transmitted qP and qSV waves take lesser values in comparison to the angles of the respective reflected waves. However, the transmitted EA wave continues to propagate along the interface. It is also observed that for qP wave incidence, there is no critical angle.

It is assumed that the roots for H_1 having positive real parts, denoted by $q_1^{(1)}, q_3^{(1)}$, and $q_5^{(1)}$ indicate the reflected qP, qSV, and EA waves, respectively, which move along the positive z direction. It is also presumed that the negative value of the root with positive real part in H_1, denoted by $q_2^{(1)}$ denotes the incident qP wave, which moves against the direction of the positive z-axis. Along the same lines, it is also assumed that the negative values of the roots with positive real parts in H_2, denoted by $q_2^{(2)}, q_4^{(2)}$, and $q_6^{(2)}$ signify the transmitted qP, qSV, and EA waves, respectively, which move against the direction of the positive z-axis.

Now, for each $q_n^{(m)}$, $n = 1, 2, \ldots, 6$, $m = 1, 2$, the expressions of $W^{(m)}$ and $\Phi^{(m)}$ may be written as

$$W_n = \Delta_1(q_n^{(m)})/\Delta(q_n^{(m)}), \quad \Phi_n = \Delta_2(q_n^{(m)})/\Delta(q_n^{(m)}), \tag{23}$$

where $\Delta_i(q_n^{(m)})$ and $\Delta(q_n^{(m)})$, for $i = 1, 2$ are explained in Appendix. With relation (23), the complete structure of the wave field in the lower half-space (H_1) may be expressed by means of the formal solutions for the mechanical displacements and electric potential as

$$\begin{aligned}
u^{(1)} &= U_\varsigma^{(1)} \exp[ik\kappa_\varsigma^{(1)}] + \sum_{n=1,3,5} U_n^{(1)} \exp[ik\kappa_n^{(1)}],\\
w^{(1)} &= W_\varsigma^{(1)} U_\varsigma^{(1)} \exp[ik\kappa_\varsigma^{(1)}] + \sum_{n=1,3,5} W_n^{(1)} U_n^{(1)} \exp[ik\kappa_n^{(1)}],\\
\phi^{(1)} &= \Phi_\varsigma^{(1)} U_\varsigma^{(1)} \exp[ik\kappa_\varsigma^{(1)}] + \sum_{n=1,3,5} \Phi_n^{(1)} U_n^{(1)} \exp[ik\kappa_n^{(1)}].
\end{aligned} \tag{24}$$

Using Eq. (24), the expressions for the mechanical stresses and electric displacement in H_1 are written as

$$\begin{aligned}
\tau_{33}^{(1)} &= ikJ_{1\varsigma}^{(1)} U_\varsigma^{(1)} \exp[ik\kappa_\varsigma^{(1)}] + \sum_{n=1,3,5} ikJ_{1n}^{(1)} U_n^{(1)} \exp[ik\kappa_n^{(1)}],\\
\tau_{13}^{(1)} &= ikJ_{2\varsigma}^{(1)} U_\varsigma^{(1)} \exp[ik\kappa_\varsigma^{(1)}] + \sum_{n=1,3,5} ikJ_{2n}^{(1)} U_n^{(1)} \exp[ik\kappa_n^{(1)}],\\
D_3^{(1)} &= ikJ_{3\varsigma}^{(1)} U_\varsigma^{(1)} \exp[ik\kappa_\varsigma^{(1)}] + \sum_{n=1,3,5} ikJ_{3n}^{(1)} U_n^{(1)} \exp[ik\kappa_n^{(1)}].
\end{aligned} \tag{25}$$

Similarly, the complete structure of the wave field in the upper half-space (H_2) may be expressed by means of the formal solution for the mechanical displacements and electric potential as

$$\begin{aligned}
u^{(2)} &= \sum_{n=2,4,6} U_n^{(2)} \exp[ik\kappa_n^{(2)}],\\
w^{(2)} &= \sum_{n=2,4,6} W_n^{(2)} U_n^{(2)} \exp[ik\kappa_n^{(2)}],\\
\phi^{(2)} &= \sum_{n=2,4,6} \Phi_n^{(2)} U_n^{(2)} \exp[ik\kappa_n^{(2)}],
\end{aligned} \tag{26}$$

using which, the stresses and electric displacement in H_2 are derived as

$$\begin{aligned} \tau_{33}^{(2)} &= \sum_{n=2,4,6} ikJ_{1n}^{(2)}U_n^{(2)}\exp[ik\kappa_n^{(2)}], \\ \tau_{13}^{(2)} &= \sum_{n=2,4,6} ikJ_{2n}^{(2)}U_n^{(2)}\exp[ik\kappa_n^{(2)}], \\ D_3^{(2)} &= \sum_{n=2,4,6} ikJ_{3n}^{(2)}U_n^{(2)}\exp[ik\kappa_n^{(2)}]. \end{aligned} \tag{27}$$

The expressions $J_{ik}^{(m)}$ $(i = 1,2,3;\ m = 1,2;\ k = \varsigma, n)$, $\kappa_\varsigma^{(1)}$ and $\kappa_n^{(m)}$ are provided in Appendix. The term ς can assume values 2 or 4 representing the incidence of qP or qSV waves, respectively. In this study, ς takes value 2 for incident qP wave. Now, on using the boundary conditions (15–20), the Eqs. (24)–(27) result in a system of equations provided in matrix form as

$$[A_{ij}]_{6\times 6}\left[\frac{U_1^{(1)}}{U_2^{(1)}}\quad \frac{U_3^{(1)}}{U_2^{(1)}}\quad \frac{U_5^{(1)}}{U_2^{(1)}}\quad \frac{U_2^{(2)}}{U_2^{(1)}}\quad \frac{U_4^{(2)}}{U_2^{(1)}}\quad \frac{U_6^{(2)}}{U_2^{(1)}}\right]^T = [B_{ij}]_{6\times 1}, \tag{28}$$

where the terms A_{ij} and B_{ij} are provided in Appendix. Using Cramer's rule, the expressions of the amplitude ratios of all reflected and transmitted waves are attained from Eq. (28), which display their dependence on various parameters *viz.* incident angle, rotation, initial stresses, and PFRC parameters. However, the amplitude ratios alone are not enough to verify if the system's entire energy remains conserved. Thus, using the amplitude ratio expressions, we derived the expressions of the energy ratios of all reflected and transmitted waves, and also the energy of interaction among them and provided graphical illustrations. In doing so, the law of the conservation of energy is automatically validated.

7. Energy Ratios

In this section, the energy shared among the incident wave and all reflected and transmitted waves is evaluated across a unit area surface element of infinitesimal thickness on the interface $(z = 0)$.

The averaged energy flux across the unit surface area, following Ref. [9], is given by

$$P = -Re[\tau_{13}\bar{\dot{u}} + \tau_{33}\bar{\dot{w}} + \tau_{13}^{\circ}(u_{,1}\bar{\dot{u}} + w_{,1}\bar{\dot{w}})$$
$$+ \tau_{33}^{\circ}(u_{,3}\bar{\dot{u}} + w_{,3}\bar{\dot{w}}) - \overline{\dot{D}_3}\phi]/2. \tag{29}$$

With the aid of Eqs. (24)–(27), Eq. (29) takes the following form:

$$P_{\alpha\beta}^{(m)} = k^2 c[J_{2\alpha}^{(m)} U_{\alpha}^{(m)} \overline{U_{\beta}^{(m)}} + J_{1\alpha}^{(m)} \overline{W_{\beta}^{(m)}} U_{\alpha}^{(m)} \overline{U_{\beta}^{(m)}}$$
$$+ (\tau_{13}^{\circ} + q_{\alpha}^{(m)} \tau_{33}^{\circ})(1 + W_{\alpha}^{(m)} \overline{W_{\beta}^{(m)}}) U_{\alpha}^{(m)} \overline{U_{\beta}^{(m)}}$$
$$+ \overline{J_{3\alpha}^{(m)}} \Phi_{\beta}^{(m)} \overline{U_{\alpha}^{(m)}} U_{\beta}^{(m)}]/2. \tag{30}$$

The general expression of the energy ratios is as follows:

$$E_{\alpha\beta}^{(m)} = P_{\alpha\beta}^{(m)} / P_{22}^{(1)}. \tag{31}$$

The energy ratios of all reflected/transmitted waves, along with the interaction energy ratio among the various waves, can be evaluated from Eq. (31). As per the notations, $P_{22}^{(1)}$ denotes the energy of incident qP wave; $E_{\alpha\alpha}^{(1)}$ ($\alpha = 1, 3$ and 5) denotes the energy ratios of the reflected qP, qSV, and EA waves, respectively; $E_{\alpha\alpha}^{(2)}$ ($\alpha = 2, 4$ and 6) denotes the energy ratios of transmitted qP, qSV, and EA waves, respectively; $E_{2\alpha}^{(1)}$ ($\alpha = 1, 3$ and 5) denotes the interacting energy among the incident wave and all reflected waves; $E_{\alpha 2}^{(1)}$ ($\alpha = 1, 3$ and 5) correspond to the interacting energy between all reflected waves and the incident wave; $E_{\alpha\beta}^{(1)}$ ($\alpha, \beta = 1, 3$ and 5 with $\alpha \neq \beta$) corresponds to the interaction energy among all reflected waves; and $E_{\alpha\beta}^{(2)}$ ($\alpha, \beta = 2, 4$ and 6 with $\alpha \neq \beta$) corresponds to the interaction energy among all transmitted waves. The expression of the complete energy of interaction considering the incident wave and all reflected and transmitted waves is as follows:

$$E_{\text{int}} = \sum_{\alpha=1,3,5} (E_{2\alpha}^{(1)} + E_{\alpha 2}^{(1)}) + \sum_{\alpha=1,3,5} \left(\sum_{\beta=1,3,5} E_{\alpha\beta}^{(1)} - E_{\alpha\alpha}^{(1)} \right)$$
$$+ \sum_{\alpha=2,4,6} \left(\sum_{\beta=2,4,6} E_{\alpha\beta}^{(2)} - E_{\alpha\alpha}^{(2)} \right), \tag{32}$$

and according to the energy conservation law, we must have

$$\sum_{\alpha=1,3,5} E^{(1)}_{\alpha\alpha} + \sum_{\alpha=2,4,6} E^{(2)}_{\alpha\alpha} + E_{\text{int}} = 1. \tag{33}$$

We have numerically evaluated the absolute values of the energy ratios of all reflected/transmitted waves, interaction energy ratios among the various waves, and the net energy of the system, and provided graphical demonstrations. It has been numerically calculated and graphically showed that the total energy is ≈ 1 for all variations of initial stresses and rotation, which shows that the net energy remains conserved in this study.

8. Special Cases of Interest

8.1. *Normal Incidence of qP Wave*

Observe the natures of all energy ratios in Figures 5(a)–(g) when the incident wave strikes the interface at $\theta = 0°$ for all cases (with and without initial stresses and rotation). Clearly, the energy carried by all the reflected waves is 0 (or ≈ 0) in this case. Considering transmitted waves, it is observed that the transmitted qP wave carries maximum energy in this case (≈ 1), while the other transmitted waves carry little to no energy. The interaction energy is also quite negligible in this case. This shows that for a normally incident qP wave, no waves are generated other than a transmitted qP wave, which propagates with the entire energy. When qSV wave is normally incident at the interface, one may obtain equivalent results.

8.2. *Grazing Incidence of qP Wave*

In this case, observe all the energy ratios again in Figures 5(a)–(g) when qP wave grazes the interface (i.e., $\theta = 90°$) for all cases (with and without initial stresses and rotation). It can be seen quite clearly that the reflected qP wave carries maximum energy in this case (≈ 1), while the magnitudes of energies of other reflected waves, all transmitted waves and interaction energy are negligible. This goes to show that for the grazing incidence of qP wave, no waves are generated except for a reflected qP wave that propagates with the

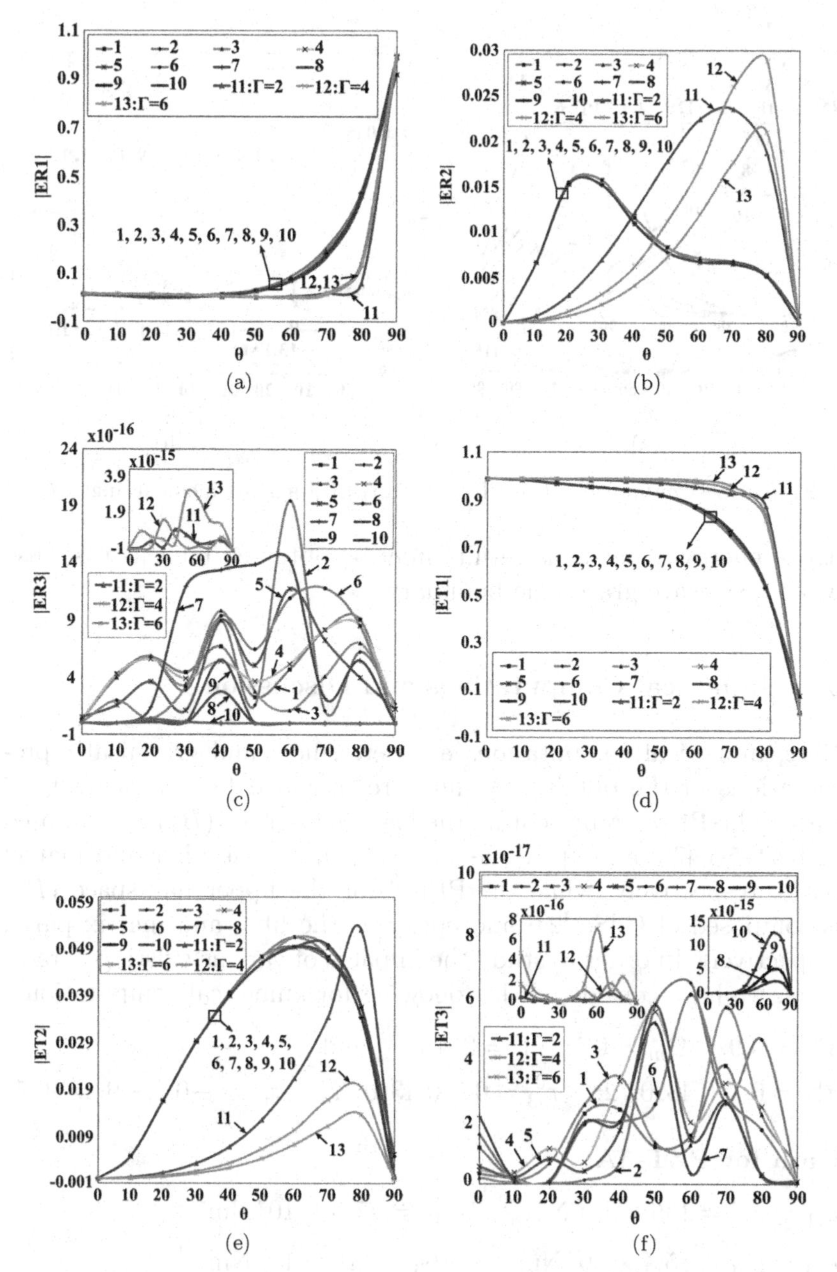

Figure 5. Absolute values of energy ratios: (a) |ER1|, (b) |ER2|, (c) |ER3|, (d) |ET1|, (e) |ET2|, (f) |ET3|.

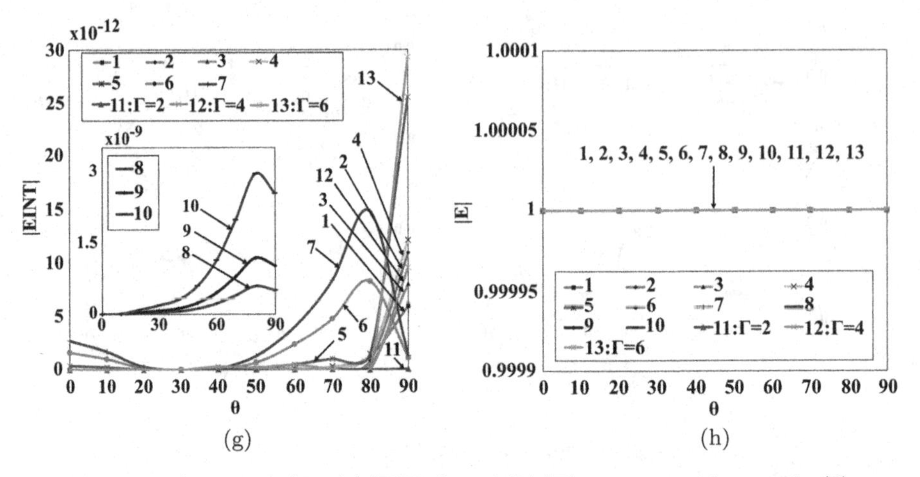

Figure 5 (*Continued*). (g) |EINT|, and (h) |E| against incident angle (θ).

entire energy. Analogous results may be obtained for this case also when qSV wave grazes the boundary.

9. Numerical Computations and Discussion

Here, numerical computations are performed, and graphical representations of the obtained results are provided for an incident qP wave. The PFRC representing the lower half-space (H_1) is composed of PZT-5A [23] and epoxy, as in Ref. [14], as the fiber and matrix parts, respectively, while the PFRC for the upper half-space (H_2) is composed of CdSe [24] and epoxy as the fiber and matrix parts, respectively. In order to study the impacts of various existing parameters, we have considered the following for numerical computations:

$$\Omega_f = 0.9, \quad \Omega_m = 0.1; \quad \Gamma = 2, 4, 6; \quad \text{and}$$
$$\tau_{11}^0 = 0.1, 0.43, 0.72; \quad \tau_{33}^0 = 0.1, 0.43, 0.72; \quad \tau_{13}^0 = -0.1, -0.2, -0.5.$$

Data for PZT-5A:

$$c_{11} = c_{22} = 139 \times 10^9\,\text{Nm}^{-2}, \quad c_{12} = 77.8 \times 10^9\,\text{Nm}^{-2},$$
$$c_{23} = c_{13} = 75.4 \times 10^9\,\text{Nm}^{-2}, \quad c_{33} = 113 \times 10^9\,\text{Nm}^{-2},$$
$$c_{44} = 25.6 \times 10^9\,\text{Nm}^{-2}, \quad e_{15} = e_{24} = 13.4\,\text{Cm}^{-2},$$

$$e_{32} = e_{31} = -6.98\,\mathrm{Cm}^{-2}, \quad e_{33} = 13.8\,\mathrm{Cm}^{-2}, \quad \rho = 7750\,\mathrm{Kgm}^{-3},$$
$$\varepsilon_{11} = \varepsilon_{22} = 60 \times 10^{-10}\,\mathrm{C^2N^{-1}m^{-2}}, \quad \varepsilon_{33} = 54.7 \times 10^{-10}\,\mathrm{C^2N^{-1}m^{-2}}.$$

Data for CdSe:

$$c_{11} = c_{22} = 74.1 \times 10^9\,\mathrm{Nm}^{-2}, \quad c_{12} = 45.2 \times 10^9\,\mathrm{Nm}^{-2},$$
$$c_{23} = c_{13} = 39.3 \times 10^9\,\mathrm{Nm}^{-2}, \quad c_{33} = 83.6 \times 10^9\,\mathrm{Nm}^{-2},$$
$$c_{44} = 13.2 \times 10^9\,\mathrm{Nm}^{-2}, \quad e_{15} = e_{24} = -0.138\,\mathrm{Cm}^{-2},$$
$$e_{31} = e_{32} = -0.16\,\mathrm{Cm}^{-2}, \quad e_{33} = 0.347\,\mathrm{Cm}^{-2}, \quad \rho = 5504\,\mathrm{Kgm}^{-3},$$
$$\varepsilon_{11} = \varepsilon_{22} = 8.26 \times 10^{-11}\,\mathrm{C^2N^{-1}m^{-2}},$$
$$\varepsilon_{33} = 9.03 \times 10^{-11}\,\mathrm{C^2N^{-1}m^{-2}}.$$

Data for Epoxy:

$$c_{11} = c_{22} = 3.86 \times 10^9\,\mathrm{Nm}^{-2}, \quad c_{12} = 2.57 \times 10^9\,\mathrm{Nm}^{-2},$$
$$c_{23} = c_{13} = 2.57 \times 10^9\,\mathrm{Nm}^{-2}, \quad c_{33} = 3.86 \times 10^9\,\mathrm{Nm}^{-2},$$
$$c_{44} = 2.57 \times 10^9\,\mathrm{Nm}^{-2}, \quad e_{ij} = 0\,\mathrm{Cm}^{-2}\ (\forall i, j),$$
$$\rho = 1210\,\mathrm{Kgm}^{-3}, \quad \varepsilon_{11} = \varepsilon_{22} = \varepsilon_{33} = 7.9 \times 10^{-11}\,\mathrm{C^2N^{-1}m^{-2}}.$$

Figures 5(a)–(h) demonstrate the nature of the absolute values of the energy ratios of reflected qP wave |ER1|, reflected qSV wave |ER2|, reflected EA wave |ER3|, transmitted qP wave |ET1|, transmitted qSV wave |ET2|, transmitted EA wave |ET3|, interaction energy |EINT|, and net energy |E|, respectively, against the incident angle $\theta \in [0^\circ, 90^\circ]$. Curve 1 in each figure represents the case without the effects of any initial stresses or rotation. Curves 2, 3, and 4 represent the variations of normal initial stress τ_{11}^0 when $\tau_{11}^0 = 0.1, 0.43$, and 0.72, respectively, for $\Gamma = 0$. Curves 5, 6, and 7 represent the variations of normal initial stress τ_{33}^0 when $\tau_{33}^0 = 0.1, 0.43$, and 0.72, respectively, for $\Gamma = 0$. Curves 8, 9, and 10 represent the variations of shear initial stress τ_{13}^0 when $\tau_{13}^0 = -0.1, -0.2$, and -0.5, respectively, for $\Gamma = 0$. Finally, curves 11, 12, and 13 represent the variations of rotation when $\Gamma = 2, 4$, and 6, respectively, when no initial stress is prevalent.

9.1. *Energy Ratio of Reflected qP Wave (|ER1|)*

From Figure 5(a), it can be seen that for all variations of normal and shear initial stresses, |ER1| starts from value 0 at $\theta = 0°$ and has minimum magnitude till $\theta = 45°$. Past that, |ER1| keeps gaining magnitude considerably and ultimately gains maximum magnitude of $\approx$1 at $\theta = 90°$. It is found that for all variations of all considered initial stresses, the natures of |ER1| are almost identical with nearly negligible visual demarcation. This is because, as qP waves are the fastest waves propagating with extremely high velocities, the initial stresses are unable to significantly impact their propagation behavior. When rotation is introduced with $\Gamma = 2$, it is observed that |ER1| assumes the lowest values in almost the entire range of the incident angle, till $\theta = 80°$, past which, |ER1| gains magnitude at a large rate and has the maximum value of $\approx$1 at $\theta = 90°$. When rotation is increased to values $\Gamma = 4$ and 6, the natures of |ER1| are almost the same as the case of $\Gamma = 2$ till $\theta = 70°$, past which, |ER1| assume a greater magnitude for both cases for the remaining angles.

9.2. *Energy Ratio of Reflected qSV Wave (|ER2|)*

In Figure 5(b), the nature of energy carried by the reflected qSV wave for variations of initial stresses and rotation is shown. In this case, it can be seen that for all variations of normal and shear initial stresses, |ER2| starts from value 0 at $\theta = 0°$, past which, its magnitude keeps increasing till it becomes maximum in the range $23° < \theta < 27°$. Beyond $\theta = 27°$, the magnitude of |ER2| starts diminishing slowly and continuously until it assumes value $\approx$0 at $\theta = 90°$. A general observation of the effects of initial stresses reveals a stark difference from the case of reflected qP wave. It is observed that although the natures of |ER2| for all variations of $(\tau_{11}^0, \tau_{33}^0, \tau_{13}^0)$ are almost similar, they are still somewhat visually clearer in this case. This is because, as qSV waves are slower than qP waves, the initial stresses have a relatively more profound effect on qSV waves in contrast to qP waves. Due to the same reason, rotation also has a more significant impact on |ER2| in contrast to |ER1|. Meticulous scrutiny reveals that for most values of θ, increasing values of rotation disfavor |ER2|, and in general, the magnitude of |ER2| is maximum more often in the presence of rotation than in its absence.

9.3. *Energy Ratio of Reflected EA Wave* (|*ER3*|)

It is evident from Figure 5(c) that the energy carried by the reflected EA wave is almost negligible. It is physically justified as per the following points.

Keep in mind that the energy ratio was calculated at a unit area surface element of infinitesimal thickness lying at the interface. The scattering bulk waves can import/export energy into/from the considered element. However, it is known that the evanescent EA waves propagate, while they carry energy along the interface of the two half-spaces. Thus, owing to the infinite thickness of the considered element, the EA waves do not make a significant contribution to the energy fluxes in and out of it. However, for a complete analysis of the problem, the energy carried by the reflected EA wave has also been analyzed meticulously. Numerical simulations show that among all considered variations of τ_{11}^0, the magnitude of |ER3| is least when $\tau_{11}^0 = 0.1, \forall\theta \in [0°, 90°]$ except for $\theta = 60°$, where it becomes maximum. The magnitudes of |ER3| for $\tau_{11}^0 = 0, 0.43$, and 0.72 take values in between. It is observed that |ER3| is encouraged for increasing values of $\tau_{33}^0, \forall\theta \in [0°, 90°]$. Finally, it is found that |ER3| has the least magnitude for all values of τ_{13}^0 among all considered cases and tends to increase with increasing magnitudes of τ_{13}^0. However, the reflected EA wave carries the least energy among all considered cases when $\tau_{13}^0 = -0.5$. When rotation is considered, it is observed that |ER3| assumes peak values for increasing values of Γ, and the magnitude of |ER3| is maximum more often in the presence of rotation than in its absence.

9.4. *Energy Ratio of Transmitted qP Wave* (|*ET1*|)

It is observed from Figure 5(d) that for all variations of initial stresses, |ET1| starts with maximum magnitude ≈1 at $\theta = 0°$ and slowly but gradually keeps losing energy until almost all of it is exhausted at $\theta = 90°$. Just like the case of |ER1|, it is found that the nature of |ET1| is almost identical with nearly negligible visual demarcation for all variations of all types of initial stresses. This can also be attributed to the fact that the qP waves are the fastest waves propagating with extremely high velocities, for which the initial stresses are unable to significantly impact the energies carried by them. It can be seen that the magnitudes of |ET1| are the greatest under the influence of rotation. For most values of θ, |ET1| is encouraged for increasing values of Γ.

9.5. *Energy Ratio of Transmitted qSV Wave (|ET2|)*

From Figure 5(e), the influences of initial stresses and rotation on the energy carried by the transmitted qSV wave is observed. In this case, it can be seen that for all variations of initial stresses, |ET2| starts from value 0 at $\theta = 0°$, past which, its magnitude keeps increasing till it becomes maximum in the range $60° < \theta < 70°$. Beyond $\theta = 70°$, the transmitted qSV wave starts losing energy rapidly until it has exhausted almost all of it and assumes value ≈ 0 at $\theta = 90°$. Just like the case of transmitted qP wave, it is observed here that although the nature of |ET2| for all variations of $(\tau_{11}^0, \tau_{33}^0, \tau_{13}^0)$ is almost similar, they are somewhat visually clearer in this case. It can be chalked up to the fact that as qSV waves are slower than qP waves, the initial stresses have a relatively more profound effect on qSV waves in contrast to qP waves. Due to the same reason, the effect of rotation is also more significant on |ET2| in contrast to |ET1|. It is found that for all values of θ, increasing values of rotation disfavor |ET2|, and in general, the magnitude of |ET2| is lesser in the presence of rotation than in its absence.

9.6. *Energy Ratio of Transmitted EA Wave (|ET3|)*

As can be observed from Figure 5(f), the energy carried by the transmitted EA wave is almost negligible. This is because, similar to the reflected EA wave, the transmitted EA wave also propagates along the interface. Thus, it makes no significant contribution to the net energy of the system. However, once again, for a complete analysis of the problem, the energy carried by the transmitted EA wave has also been analyzed meticulously. Numerical simulations reveal that the magnitudes of energy carried by the transmitted EA wave are the least in the presence of τ_{11}^0 and τ_{33}^0. On the other hand, a significant impact of τ_{13}^0 is observed. The magnitude of |ET3| is found to attain the largest values among all considered cases, with increasing magnitudes of τ_{13}^0 encouraging |ET3|. When rotation is considered, it is observed that the magnitude of energy carried by the transmitted EA wave increases for increasing values of Γ.

9.7. *Interaction Energy Ratio* (|*EINT*|)

From Figure 5(g), it is observed that the interaction energy ratio has a very low magnitude for all variations of initial stresses and rotation, in general. The magnitude of |EINT| is found to be almost zero in the range $0° < \theta < 80°$ for all values of τ_{11}^{0}. Beyond $\theta = 80°$, |EINT| is encouraged by increasing values of τ_{11}^{0}. It is found that τ_{33}^{0} has a more marked effect on |EINT| compared to τ_{11}^{0}, for increasing values of τ_{33}^{0}, |EINT| keeps increasing $\forall\theta$, and eventually surpasses the magnitude of interacting energy ratio for all variations of τ_{11}^{0}. Considering shear stress τ_{13}^{0}, it can be seen that |EINT| assumes the largest values among all considered cases with increasing values for increasing magnitudes of τ_{13}^{0}. With the introduction of rotation, it is observed that in the range $0° < \theta < 80°$, |EINT| assumes the least value among all considered cases for $\Gamma = 2$. However, beyond $\theta = 80°$, increasing values of rotation are found to encourage |EINT|.

9.8. *Net Energy Ratio* (|*E*|)

The net energy ratio (which is the sum of the energy ratios of all reflected waves, transmitted waves, and interacting energy ratio) for all cases is demonstrated in Figure 5(h). The net energy has been numerically calculated up to five decimal places to derive accurate results. It is seen that for all values of considered initial stresses and rotation, the net energy takes a constant value of one $\forall\theta \in [0°, 90°]$. It shows that for any variation of initial stresses or rotation, the net energy in the system remains conserved. This acts as a validation of the present problem since the law of conservation of energy holds good in this case.

10. Conclusions

In this work, the micro-mechanics model of PFRC is established by using SM and RM techniques, and some of the electro-mechanical advantages of PFRC over monolithic piezoelectric materials are graphically shown. Thereafter, the reflection and transmission characteristics of plane waves in an initial-stressed rotating structure comprised of two dissimilar PFRC half-spaces are analyzed. The PFRC considered for the two half-spaces are comprised of PZT-5A fiber-epoxy combination and CdSe fiber-epoxy combination. The inclusion of shear and normal initial stresses and rotation are considered for making the present study more realistic to the actual phenomenon. The amplitude ratios of all reflected and transmitted waves are derived, using which the expressions of energy ratios of reflected waves, transmitted waves, and interaction energy are evaluated and illustrated graphically. The salient features of this work are encapsulated as follows:

(i) The data for the two dissimilar PFRCs considered for both half-spaces are derived using the SM and RM techniques. These techniques rely only on the material properties of the constituent fibers and matrix and work with some simple assumptions. The natures of most of the electro-mechanical data derived by using micro-mechanics model bear a striking resemblance to the same obtained by other analytical and numerical techniques, such as continuum mechanics, finite-element method, asymptotic homogenization method, and method of cells approach. After deriving the data of both media for all values of $\Omega_f \in [0, 1]$, we chose the specific values of the fiber and matrix volume fractions as 0.9 and 0.1, respectively. This is because, for those specific values, both PFRCs attained enhanced electro-mechanical properties in comparison to their monolithic PE fiber parts. The well-established SM and RM techniques used here can be applied in a wide range of scientific and engineering fields that require the development and utilization of smart composites, such as piezo-thermoelastic fiber-reinforced composites, piezoelectric fiber-reinforced composites, and piezomagnetic fiber-reinforced composites.

(ii) No critical angle is observed for the incidence of qP wave in the lower PFRC half-space (PZT-5A-epoxy combination).

(iii) It is found that increasing magnitudes of normal initial stresses $(\tau_{11}^0, \tau_{33}^0)$ and shear initial stress (τ_{13}^0) do not cause a significant change in the amount of energies carried by the reflected and transmitted qP waves, respectively. However, the differences in energies carried by the reflected and transmitted qSV waves are more prominent than those of the qP waves. These results can be attributed to the fact that qP waves are the fastest waves, and thus, the effects of varying magnitudes of normal initial stresses are minimal. qSV waves being somewhat slower in comparison experience the effects more. Influences of initial stresses are also studied for EA waves, which are surface waves propagating along the interface. However, their contribution to the net energy of the system is negligible. The presence of different initial stresses causes compaction and rarefaction of the structure in particular directions; this causes the reflected and transmitted waves to carry different amounts of energies, which has been physically justified. The analyzed results on initial stresses may be utilized in an array of scientific and engineering disciplines that involve thermal/chemical influences, deliberate application of pre-stress, materials with non-uniform properties, etc.

(iv) It is observed that varying magnitudes of the rotation parameter Γ do not cause a significant change in the amount of energies carried by the reflected and transmitted qP waves. The influence of rotation is more evident for the reflected and transmitted qSV waves. The effects of Γ on the reflected and transmitted EA waves are also studied, although their contribution to the net energy of the system is negligible. Here also, it can be concluded that as the velocities of qP and qSV waves keep decreasing, the rotation parameter keeps becoming more dominant. Also, as the entire structure is rotating about the z-axis and EA waves are surface waves that are propagating along the interface (x-axis), the effects of rotation on these waves are quite distinguishable. The results obtained on analyzing rotation may be utilized in applications involving gyroscopes that are required to measure the angular velocities of rotating bodies.

(v) Special cases of normal incidence ($\theta = 0°$) and grazing incidence ($\theta = 90°$) of qP wave are discussed.

Considering normal incidence, it is seen that the transmitted qP wave carries maximum energy and all other waves carry negligible amounts of energy for all variations of initial stresses and rotation.

For grazing incidence, it is seen that the reflected qP wave carries the maximum amount of energy, while the other waves carry meager amounts of energy for all variations of considered initial stresses and rotation.

Analogous outcomes may be achieved for both normal incidence and grazing incidence of qSV wave.

(vi) It is observed that the net energy (sum of the energy ratios of all reflected waves, all transmitted waves, and interaction energy) assumes value ≈1 for all values of considered initial stresses and rotation. This shows that no loss of energy occurs in the present study and the law of conservation of energy is therefore satisfied.

References

[1] Achenbach, J. D. (1973). *Wave Propagation in Elastic Solids*, American Elsevier Pub. Co.

[2] Li, Y. and Wei, P. (2016). Reflection and transmission through a microstructured slab sandwiched by two half-spaces. *European Journal of Mechanics — A Solids*, 57, 1–17.

[3] Nayfeh, A. H. and Anderson, M. J. (2002). Wave propagation in layered anisotropic media with applications to composites. *Journal of the Acoustical Society of America*, 108(2), 471–472.

[4] Curie, J. and Curie, P. (1880). Développement par compression de l'électricité polaire dans les cristaux hémièdres à faces inclinées. *Bulletin de Minéralogie*, 3(4), 90–93.

[5] Yuan, X. and Zhu, Z. H. (2012). Reflection and refraction of plane waves at interface between two piezoelectric media. *Acta Mechanica*, 223(12), 2509–2521.

[6] Pang, Y., Wang, Y. S., Liu, J. X., and Fang, D. N. (2008). Reflection and refraction of plane waves at the interface between piezoelectric and piezomagnetic media. *International Journal of Engineering Science*, 46(11), 1098–1110.

[7] Guha, S., Singh, A. K., and Das, A. (2019). Analysis on different types of imperfect interfaces between two dissimilar piezothermoelastic half-spaces on reflection and refraction phenomenon of plane waves. *Waves Random Complex Media*, 31(4), 1–30.

[8] Biot, M. A. (1940). The influence of initial stress on elastic waves. *Journal of Applied Physics*, 11(8), 522–530.

[9] Guo, X. and Wei, P. (2014). Effects of initial stress on the reflection and transmission waves at the interface between two piezoelectric half spaces. *International Journal of Solids and Structures*, 51(21–22), 3735–3751.

[10] Fang, H. Y., Yang, J. S., and Jiang, Q. (2002). Rotation sensitivity of waves propagating in a rotating piezoelectric plate. *International Journal of Solids and Structures*, 39(20), 5241–5251.

[11] Yuan, X., Jiang, Q., and Yang, F. (2016). Wave reflection and transmission in rotating and stressed pyroelectric half-planes. *Applied Mathematics and Computation*, 289, 281–297.

[12] Ray, M. C. (2006). Micromechanics of piezoelectric composites with improved effective piezoelectric constant. *International Journal of Mechanics and Materials in Design*, 3(4), 361–371.

[13] Berger, H., Kari, S., Gabbert, U., Rodriguez-Ramos, R., Guinovart, R., Otero, J. A., and Bravo-Castillero, J. (2005). An analytical and numerical approach for calculating effective material coefficients of piezoelectric fiber composites. *International Journal of Solids and Structures*, 42(21–22), 5692–5714.

[14] Kumar, A. and Chakraborty, D. (2009). Effective properties of thermo-electro-mechanically coupled piezoelectric fiber reinforced composites. *Materials & Design*, 30(4), 1216–1222.

[15] Hill, R. (1964). Theory of mechanical properties of fibre-strengthened materials: I. Elastic behaviour. *Journal of the Mechanics and Physics of Solids*, 12(4), 199–212.

[16] Samal, S. K. and Chattaraj, R. (2011). Surface wave propagation in fiber-reinforced anisotropic elastic layer between liquid saturated porous half space and uniform liquid layer. *Acta Geophysica*, 59(3), 470–482.

[17] Bisheh, H., Wu, N., and Hui, D. (2019). Polarization effects on wave propagation characteristics of piezoelectric coupled laminated fiber-reinforced composite cylindrical shells. *International Journal of Mechanical Sciences*, 161–162, 105028.

[18] Singh, A. K. and Guha, S. (2020). Reflection of plane waves from the surface of a piezothermoelastic fiber-reinforced composite half-space. *Mechanics of Advanced Materials and Structures*, 1–13.

[19] Guha, S. and Singh, A. K. (2020). Effects of initial stresses on reflection phenomenon of plane waves at the free surface of a rotating piezothermoelastic fiber-reinforced composite. *International Journal of Mechanical Sciences*, 181, 105766.

[20] Benveniste, Y. and Dvorak, G. J. (1992). Uniform fields and universal relations in piezoelectric composites. *Journal of the Mechanics and Physics of Solids*, 40(6), 1295–1312.

[21] Xia, X. K. and Shen, H. S. (2009). Nonlinear vibration and dynamic response of FGM plates with piezoelectric fiber reinforced composite actuators. *Composite Structures*, 90(2), 254–262.

[22] Auriault, J. L. (2004). Body wave propagation in rotating elastic media. *Mechanics Research Communications*, 31(1), 21–27.

[23] Alshaikh, F. A. (2012). Reflection of Quasi vertical transverse waves in the thermo-piezoelectric material under initial stress (Green-Lindsay Model). *International Journal of Pure and Applied Sciences and Technology*, 13(1), 27–39.
[24] Sharma, J. N., Walia, V., and Gupta, S. K. (2008). Reflection of piezothermoelastic waves from the charge and stress free boundary of a transversely isotropic half space. *International Journal of Engineering Science*, 46(2), 131–146.

Appendix A. Mathematical Expressions

$$G_1 = c_{13}^f - c_{13}^m, \quad G_2 = c_{12}^f - c_{12}^m,$$
$$G_3 = \Omega_m c_{23}^f + \Omega_f c_{23}^m, \quad G_4 = \Omega_m c_{33}^f + \Omega_f c_{33}^m,$$
$$G_5 = \Omega_m c_{22}^f + \Omega_f c_{22}^m, \quad G_6 = \Omega_m c_{55}^f + \Omega_f c_{55}^m,$$
$$G_7 = \Omega_m c_{44}^f + \Omega_f c_{44}^m, \quad P_1 = \Omega_m G_1 G_3 - \Omega_m G_2 G_4,$$
$$P_2 = c_{22}^m G_4 - c_{23}^m G_3, \quad P_3 = c_{23}^m G_4 - c_{33}^m G_3,$$
$$P_4 = \Omega_m e_{32}^f G_4 - \Omega_m e_{33}^f G_3, \quad P_5 = -\Omega_m e_{31}^f G_3 - \Omega_m e_{33}^f G_5,$$
$$P_6 = -\Omega_m G_1 G_5 + \Omega_m G_2 G_3, \quad P_7 = -c_{22}^m G_3 + c_{23}^m G_5,$$
$$P_8 = -c_{23}^m G_3 + c_{33}^m G_5, \quad \Lambda_1 = \Omega_f G_2,$$
$$\Lambda_2 = \Omega_f G_1, \quad \varphi = G_4 G_5 - G_3^2,$$
$$c_{11} = \Omega_f c_{11}^f + \Omega_m c_{11}^m + (P_1 \Lambda_1 + P_6 \Lambda_2)/\varphi,$$
$$c_{22} = (P_2 c_{22}^f + P_7 c_{23}^f)/\varphi, \quad c_{23} = (P_3 c_{22}^f + P_8 c_{23}^f)/\varphi,$$
$$c_{12} = c_{12}^m + (P_2 \Lambda_1 + P_7 \Lambda_2)/\varphi, \quad c_{13} = c_{13}^f + (P_1 c_{23}^f + P_6 c_{33}^f)/\varphi,$$
$$c_{33} = (P_3 c_{23}^f + P_8 c_{33}^f)/\varphi, \quad c_{44} = c_{44}^f c_{44}^m / G_7, \quad c_{55} = c_{55}^f c_{55}^m / G_6,$$
$$e_{31} = \Omega_f e_{31}^f - (P_4 \Lambda_1 + P_9 \Lambda_2)/\varphi,$$
$$e_{32} = e_{32}^f - (P_4 c_{22}^f + P_5 c_{23}^f)/\varphi, \; e_{33} = e_{33}^f - (P_4 c_{23}^f + P_5 c_{33}^f)/\varphi,$$
$$e_{15} = e_{15}^f [1 - (\Omega_m c_{55}^f)/G_6], \quad e_{24} = e_{24}^f [1 - (\Omega_m c_{44}^f)/G_7],$$
$$\varepsilon_{11} = \Omega_f \varepsilon_{11}^f + \Omega_m \varepsilon_{11}^m + (\Omega_f \Omega_m (e_{15}^f)^2)/G_6,$$
$$\varepsilon_{33} = \Omega_f \varepsilon_{33}^f + \Omega_m \varepsilon_{33}^m + \Omega_f (P_4 e_{32}^f + P_5 e_{33}^f)/\varphi,$$

$$\begin{aligned}
\Delta_1(q_n^{(m)}) &= (-c_{11}^{(m)}e_{15}^{(m)} + c^2 e_{15}^{(m)}\rho^{(m)}(1+\Gamma^2) - e_{15}^{(m)}\tau_{11}^0) \\
&\quad - 2e_{15}^{(m)}\tau_{13}^0 q_n^{(m)} + [c_{13}^{(m)}e_{15}^{(m)} + c_{13}^{(m)}e_{31}^{(m)} + c_{44}^{(m)}e_{31}^{(m)} \\
&\quad - c_{11}^{(m)}e_{33}^{(m)} + c^2 e_{33}^{(m)}\rho^{(m)}(1+\Gamma^2) - (e_{33}^{(m)}\tau_{11}^0 \\
&\quad + e_{15}^{(m)}\tau_{33}^0)](q_n^{(m)})^2 - 2e_{33}^{(m)}\tau_{13}^0(q_n^{(m)})^3 \\
&\quad - (c_{44}^{(m)}e_{33}^{(m)} + e_{33}^{(m)}\tau_{33}^0)(q_n^{(m)})^4, \\
\Delta_2(q_n^{(m)}) &= c_{11}^{(m)}c_{44}^{(m)} - c^2 c_{11}^{(m)}\rho^{(m)} - c^2 c_{44}^{(m)}\rho^{(m)}(1+\Gamma^2) \\
&\quad + c^4(\rho^{(m)})^2(1+\Gamma^2) + (c_{11}^{(m)} + c_{44}^{(m)})\tau_{11}^0 + (\tau_{11}^0)^2 \\
&\quad - 2c^2\rho^{(m)}\tau_{11}^0 - c^2\rho^{(m)}\tau_{11}^0\Gamma^2 \\
&\quad + (2c_{11}^{(m)}\tau_{13}^0 + 2c_{44}^{(m)}\tau_{13}^0 + 4\tau_{11}^0\tau_{13}^0 - 4c^2\rho^{(m)}\tau_{13}^0 \\
&\quad - 2c^2\Gamma^2\rho^{(m)}\tau_{13}^0)q_n^{(m)} + [c_{11}^{(m)}c_{33}^{(m)} \\
&\quad - (c_{13}^{(m)})^2 - 2c_{13}^{(m)}c_{44}^{(m)} - (c_{33}^{(m)} + c_{44}^{(m)})c^2\rho^{(m)} \\
&\quad + (c_{33}^{(m)} + c_{44}^{(m)})\tau_{11}^0 + 4(\tau_{13}^0)^2 + (c_{11}^{(m)} + c_{44}^{(m)})\tau_{33}^0 + 2\tau_{11}^0\tau_{33}^0 \\
&\quad - c^2 c_{33}^{(m)}\rho^{(m)}\Gamma^2 - 2c^2\tau_{33}^0\rho^{(m)} - c^2\rho^{(m)}\Gamma^2\tau_{33}^0](q_n^{(m)})^2 \\
&\quad + (2(c_{33}^{(m)} + c_{44}^{(m)})\tau_{13}^0 + 4\tau_{13}^0\tau_{33}^0)(q_n^{(m)})^3 + (c_{33}^{(m)}c_{44}^{(m)} \\
&\quad + (c_{33}^{(m)} + c_{44}^{(m)})\tau_{33}^0 + (\tau_{33}^0)^2)(q_n^{(m)})^4, \\
\Delta(q_n^{(m)}) &= (c_{13}^{(m)}e_{15}^{(m)} - c_{44}^{(m)}e_{31}^{(m)} + (c^2\rho^{(m)} - \tau_{11}^0)(e_{15}^{(m)} + e_{31}^{(m)}))q_n^{(m)} \\
&\quad - 2(e_{15}^{(m)} + e_{31}^{(m)})\tau_{13}^0(q_n^{(m)})^2 + (e_{33}^{(m)}(c_{13}^{(m)} + c_{44}^{(m)}) \\
&\quad - (e_{15}^{(m)} + e_{31}^{(m)})(c_{33}^{(m)} + \tau_{33}^0))(q_n^{(m)})^3, \\
J_{1k}^{(m)} &= c_{13}^{(m)} + c_{33}^{(m)}q_k^{(m)}W_k^{(m)} + e_{33}^{(m)}q_k^{(m)}\Phi_k^{(m)}, \\
J_{2k}^{(m)} &= c_{44}^{(m)}(q_k^{(m)} + W_k^{(m)}) + e_{15}^{(m)}\Phi_k^{(m)}, \\
J_{3k}^{(m)} &= e_{31}^{(m)} + e_{33}^{(m)}q_k^{(m)}W_k^{(m)} - \varepsilon_{33}^{(m)}q_k^{(m)}\Phi_k^{(m)},
\end{aligned}$$

$$\kappa_{\zeta}^{(1)} = x + q_{\zeta}^{(1)} z - ct, \qquad \kappa_{n}^{(m)} = x + q_{n}^{(m)} z - ct,$$

$$A_{11} = -1, \quad A_{12} = -1, \quad A_{13} = -1,$$

$$A_{14} = 1, \quad A_{15} = 1, \quad A_{16} = 1,$$

$$A_{21} = -W_1^{(1)}, \quad A_{22} = -W_3^{(1)}, \quad A_{23} = -W_5^{(1)},$$

$$A_{24} = W_2^{(2)}, \quad A_{25} = W_4^{(2)}, \quad A_{26} = W_6^{(2)},$$

$$A_{31} = -(J_{21}^{(1)} + \tau_{13}^0 + \tau_{33}^0 q_1^{(1)}), \quad A_{32} = -(J_{23}^{(1)} + \tau_{13}^0 + \tau_{33}^0 q_3^{(1)}),$$

$$A_{33} = -(J_{25}^{(1)} + \tau_{13}^0 + \tau_{33}^0 q_5^{(1)}), \quad A_{34} = J_{22}^{(2)} + \tau_{13}^0 + \tau_{33}^0 q_2^{(2)},$$

$$A_{35} = J_{24}^{(2)} + \tau_{13}^0 + \tau_{33}^0 q_4^{(2)}, \qquad A_{36} = J_{26}^{(2)} + \tau_{13}^0 + \tau_{33}^0 q_6^{(2)},$$

$$A_{41} = -J_{11}^{(1)} - (\tau_{13}^0 + \tau_{33}^0 q_1^{(1)}) W_1^{(1)},$$

$$A_{42} = -J_{13}^{(1)} - (\tau_{13}^0 + \tau_{33}^0 q_3^{(1)}) W_3^{(1)},$$

$$A_{43} = -J_{15}^{(1)} - (\tau_{13}^0 + \tau_{33}^0 q_5^{(1)}) W_5^{(1)},$$

$$A_{44} = J_{12}^{(2)} + (\tau_{13}^0 + \tau_{33}^0 q_2^{(2)}) W_2^{(2)},$$

$$A_{45} = J_{14}^{(2)} + (\tau_{13}^0 + \tau_{33}^0 q_4^{(2)}) W_4^{(2)},$$

$$A_{46} = J_{16}^{(2)} + (\tau_{13}^0 + \tau_{33}^0 q_6^{(2)}) W_6^{(2)},$$

$$A_{51} = -\Phi_1^{(1)}, \quad A_{52} = -\Phi_3^{(1)}, \quad A_{53} = -\Phi_5^{(1)},$$

$$A_{54} = \Phi_2^{(2)}, \quad A_{55} = \Phi_{42}^{(2)}, \quad A_{56} = \Phi_{62}^{(2)},$$

$$A_{61} = -J_{31}^{(1)}, \quad A_{62} = -J_{33}^{(1)}, \quad A_{63} = -J_{35}^{(1)},$$

$$A_{64} = J_{32}^{(2)}, \quad A_{65} = J_{34}^{(2)}, \quad A_{66} = J_{36}^{(2)},$$

$$B_{11} = 1, \quad B_{21} = W_2^{(1)}, \quad B_{31} = J_{22}^{(1)} + \tau_{13}^0 + \tau_{33}^0 q_2^{(1)},$$

$$B_{41} = J_{12}^{(1)} + (\tau_{13}^0 + \tau_{33}^0 q_2^{(1)}) W_2^{(1)}, \quad B_{51} = \Phi_2^{(1)}, \quad B_{61} = J_{32}^{(1)}.$$

https://doi.org/10.1142/9789811245367_0007

Chapter 7

Analyzing Shocks and Traveling Waves in Non-Newtonian Viscoelastic Fluids

Bikash Sahoo* and Sradharam Swain

Department of Mathematics,
National Institute of Technology Rourkela,
Odisha, India
* *bikashsahoo@nitrkl.ac.in*

Abstract

Shock waves are plane discontinuities, pulse-like in nature, and are sometimes more appropriately called shock fronts. Formation and characteristics of shock waves can be easily understood by studying the flow of a compressible fluid through a nozzle. One can see that in some cases, continuity considerations lead to the formation of plane shock waves across which there are discontinuities in pressure, density, temperature, etc. The principles of conservation of mass, momentum, and energy still apply across these plane discontinuities, but there is an entropy gain. In this chapter, we mainly discuss the recent developments of shock waves, shock-wave solutions, and traveling-wave solutions in viscoelastic generalization of Burgers' equation. We also discuss the shock wave solutions of differential equations arising due to flow of some viscoelastic fluids of differential type.

Keywords: Shock waves, traveling waves, non-Newtonian fluids, viscoelastic fluids, third grade fluid, fourth grade fluid

1. Introduction

The phenomenon of shock waves is commonly associated with aerospace engineering, in particular with supersonic flight. If an aircraft travels at supersonic speed (more than the speed of sound), then one can see formation of a shock wave around the body. Some of the features associated with the shock waves are dissipation of energy and sudden changes in velocity, pressure, and temperature. Shock waves are nonlinear waves that propagate at supersonic speed. Such disturbances occur during explosion, earthquakes, hydraulic jumps, lightning strokes, and in transonic and supersonic flows. Any sudden release of energy results in the formation of shock waves. Usually, the dissipation of mechanical, nuclear, or electrical energy in a limited space results in the formation of shock waves. These waves do not form or propagate in vacuum because their dissipative nature needs a medium. Shock waves are differentiated as strong or weak depending on the instantaneous changes in flow properties, such as temperature and pressure, that they bring about in the medium of propagation. Briefly, a shock wave can be visualized as a very thin and sharp wave front across which velocity, temperature, pressure, density, and entropy of the flow change abruptly. The modern concise definition of shock wave was first given in 1905 by the Hungarian physicist, G. Zemplen from the University of Budapest. In a fluid flow, the occurrence of shock wave is characterized by sudden changes in pressure, velocity, and temperature. Thus, the shock layer is a region of high pressure, temperature, and density compared to the freestream-flow conditions. Mathematically speaking, whenever a fluid streamline crosses a shock wave, an instantaneous increase in temperature, density, and static pressure are observed with significant decrease in the velocity of the flow. Thus, a sudden change in the medium of propagation is one of the unique features that characterizes a shock wave. These waves are regarded as discontinuities with highly localized irreversibility because of the sudden changes in the flow properties. Despite abundant research, the shock-wave phenomenon is still a mystery. Shock waves have numerous applications in various fields, such as aerodynamics, medicine, process engineering, agriculture, and biology. The ability of shock waves to abruptly increase the pressure and temperature in a medium of propagation enables their use for numerous industrial and practical applications, such as gene transfer, preservative injection into wood slats, oil extraction, drug delivery, and metal forming.

In this chapter, we mainly discuss the recent developments of shock waves, shock-wave solutions, and traveling-wave solutions in viscoelastic generalization of Burgers' equation. We also discuss the shock-wave solutions of differential equations arising due to flow of some viscoelastic fluids of differential type. The Burgers' equation is given by Ref. [1]

$$u_t + uu_x = \epsilon u_{xx}. \tag{1}$$

It is probably the simplest model that couples the dissipative viscous behavior with the nonlinear convective behavior. Its inviscid counterpart is given by

$$u_t + uu_x = 0. \tag{2}$$

This equation is related to many important topics in nonlinear partial-differential equations, such as traveling waves, shock waves, similarity solutions, singular perturbation, and numerical methods for parabolic- and hyperbolic-differential equations. We will discuss how the addition of viscoelasticity affects traveling-wave solutions of Burgers' equation.

Literature survey indicates that some of the boundary layer–flow problems of non-Newtonian viscoelastic fluids (fluids of differential type) show shock-wave behavior. The general constitutive equation of these fluids is given by

$$\mathbf{T} = -p\mathbf{I} + \sum_{j=1}^{n} \mathbf{S}_j, \quad \text{with } n = 4. \tag{3}$$

Here, p is the pressure and $\mathbf{I}$ is the identity tensor. The Rivlin–Ericksen [2] tensors $\mathbf{S}_j$ are given by

$$\begin{aligned}
\mathbf{S}_1 &= \mu\mathbf{A}_1, \\
\mathbf{S}_2 &= \alpha_1\mathbf{A}_2 + \alpha_2\mathbf{A}_1^2, \\
\mathbf{S}_3 &= \beta_1\mathbf{A}_3 + \beta_2(\mathbf{A}_1\mathbf{A}_2 + \mathbf{A}_2\mathbf{A}_1) + \beta_3(tr\mathbf{A}_1^2)\mathbf{A}_1, \\
\mathbf{S}_4 &= \gamma_1\mathbf{A}_4 + \gamma_2(\mathbf{A}_3\mathbf{A}_1 + \mathbf{A}_1\mathbf{A}_3) + \gamma_3\mathbf{A}_2^2 + \gamma_4(\mathbf{A}_2\mathbf{A}_1^2 + \mathbf{A}_1^2\mathbf{A}_2) \\
&\quad + \gamma_5(tr\mathbf{A}_2)\mathbf{A}_2 + \gamma_6(tr\mathbf{A}_2)\mathbf{A}_1^2 + [\gamma_7 tr\mathbf{A}_3 + \gamma_8 tr(\mathbf{A}_2\mathbf{A}_1)]\mathbf{A}_1.
\end{aligned} \tag{4}$$

Here, μ is the dynamic viscosity and α_i, β_i, and γ_i are the material constants. The Rivlin–Ericksen tensors $\mathbf{A}_i$ are defined by

$$\mathbf{A}_1 = (\mathrm{grad}\,\mathbf{V}) + \mathrm{grad}\,\mathbf{V})^T, \tag{5}$$

$$\mathbf{A}_n = \frac{d\mathbf{A}_{n-1}}{dt} + \mathbf{A}_{n-1}(\mathrm{grad}\,\mathbf{V}) + (\mathrm{grad}\,\mathbf{V})^T\mathbf{A}_{n-1}. \tag{6}$$

2. Shock Structure in Non-Newtonian Viscoelastic Fluids

In solid mechanics, the internal friction in a material does not preclude the velocity discontinuities in it. In linear-viscoelasticity theory, the viscous dissipation makes the externally imposed shocks faded. But it does not immediately smooth them into continuous transitions. For ideal fluids, velocity discontinuities occur when the Mach number exceeds unity. In case of Newtonian viscous fluids, such discontinuities are replaced by sudden but smooth transitions. Following Pipkin [3], in this section, we will briefly discuss the shock structure in the one-dimensional steady motion of a simple viscoelastic-fluid model. In case of viscous Newtonian fluids, an instantaneous finite change in strain requires an infinite stress, and the speed of small shearing disturbances is infinite. However, for the considered viscoelastic model, an opposite trend is observed. For the considered problem, we have two relevant sound speeds. We consider the usual sound speed when the viscoelastic effects are neglected, and in this case, the shock Mach number is denoted by M. In presence of viscoelastic terms, there is a larger but finite sound speed. If M_v denotes the viscoelastic Mach number, then obviously, $M_v < M$. For the limiting case of Newtonian viscosity, $M_v = 0$, we will show that if $M > 1$ but $M_v < 1$, the transition from one region of uniform flow to another is smooth. If $M > 1$ but $M_v = 1$, the velocity is continuous but a discontinuity in acceleration exists. A true velocity discontinuity occurs when $M_v > 1$.

The constitutive equation of the viscoelastic fluid we have considered is [2]

$$\sigma_{ij}(t) = -p[\rho(t)]\delta_{ij} + \int_{-\infty}^{t} \{\delta_{ij}K_1[t-\tau,\rho(t)]\partial G_{kk}(\tau,t)/\partial\tau \\ + K_2[t-\tau,\rho(t)]\partial G_{ij}(\tau,t)/\partial\tau\}\,d\tau, \tag{7}$$

where the strain components are defined by

$$G_{ij}(\tau,t) = \frac{\partial x_k}{\partial x_i}\frac{\partial x_k}{\partial x_j}. \tag{8}$$

In the Taylor series expansion,

$$\frac{\partial G(\tau,t)}{\partial \tau} = \sum_{n=0}^{\infty} \frac{1}{n!}(\tau - t)^n \mathbf{A}_{n+1}(t), \tag{9}$$

$\mathbf{A}_n$ are the Rivlin–Ericksen kinematic matrices. For the problem under consideration, the flow is steady and each particle moves in the x_1 direction. The flow approaches a uniform velocity, say V_0, as $x_1 \to -\infty$. Let $L(\tau)$ denote the length of such an element at time τ per unit of its length at $\tau = -\infty$. Clearly $G(\tau,t)$ is diagonal with elements $\frac{L^2(\tau)}{L^2(t)}$, 1, 1. In the limit of slow, steady motion, Eq. (7) becomes

$$\sigma_{ij} = -p(\rho)\delta_{ij} + \delta_{ij}(\lambda/2)A^{(1)}_{kk} + \mu A^{(1)}_{ij}. \tag{10}$$

The apparent viscosity λ and μ are defined by

$$\lambda/2 = \int_0^\infty K_1(\tau)\,d\tau, \quad \mu = \int_0^\infty K_2(\tau)\,d\tau. \tag{11}$$

The relaxation time T is defined by

$$T(\lambda/2 + \mu) = \int_0^\infty [K_1 + K_2]\tau\,d\tau. \tag{12}$$

For the steady one-dimensional motion under consideration, as $t \to -\infty$, $v \to v_0$ and $\rho(t) \to \rho_0$. The equation of continuity implies

$$L(t) = \rho_0/rho = v(t)/v_0. \tag{13}$$

The momentum equation yields

$$\rho_0 v_0[v(t) - v_0] = \sigma(t) - \sigma_0, \tag{14}$$

where $\sigma = \sigma_{11}$ and $\lim_{t\to-\infty}\sigma = \sigma_0$. Using the constitutive Eq. (7), we obtain

$$\sigma(\tau) = -p[\rho_0/L(t)] + \int_{-\infty}^{t} K(t-\tau)(\partial/\partial\tau)[L^2(\tau)/L^2(t)]\,d\tau, \tag{15}$$

where $K(\tau) = K_1(\tau) + K_2(\tau)$. Using some of the aforementioned results, one finally obtains

$$\rho_0 v_0^2[L(t) - 1] + p[\rho_0/L(t)] - p_0$$
$$= L^{-2}(t) \int_{-\infty}^{t} K(t-\tau)[\partial L^2(\tau)/\partial\tau]\, d\tau. \tag{16}$$

With the help of Eq. (16), now one can show that viscoelastic Mach number is always smaller than the gas dynamic Mach number. To analyze the propagation of infinitesimal discontinuity in a medium initially at rest, we let

$$L(t) = \begin{cases} 1; & \text{when } t < 0, \\ L_1; & \text{when } t = 0. \end{cases}$$

Using Eq. (16) at time $t = 0+$, one obtains

$$\rho_0 v_0^2 L_1^2(L_1 - 1) + L_1^2[p(\rho_0/L_1) - p_0] = K(0)(L_1^2 - 1). \tag{17}$$

In the limit of an infinitesimal discontinuity $L_1 - 1$, the above jump condition gives

$$\rho_0 v_0^2 - \rho_0 p'(\rho_0) = 2K(0). \tag{18}$$

For an ideal fluid, $K(0) = 0$ and the sound speed is given by $c^2 = p'(\rho_0)$. Based on this sound speed, the Mach number is given by

$$M^2 = v_0^2/p'(\rho_0). \tag{19}$$

With $K(0) \neq 0$, we get from Eq. (18) the viscoelastic Mach number

$$M_v^2 = v_0^2[p'(\rho_0) + 2K(0/\rho_0)]. \tag{20}$$

Thus, the viscoelastic Mach number is always smaller than the gas dynamic Mach number.

Again, following Pipkin [3], if we consider the special model with

$$p(t) = (p_0/\rho_0)\rho(t) = p_0/L(t), \quad K(\tau) = K_0 exp(-\tau/T), \tag{21}$$

the Mach numbers M and M_v will be given by

$$M^2 = \rho_0 v_0^2/p_0, \quad M_v^2 \rho_0 v_0^2/(p_0 + 2K_0). \tag{22}$$

Equation (16) can be converted to a first-order ordinary-differential equation. Apart from the trivial solution $L(t) = 1$, this equation has a solution,

$$(t - t_0)/T' = [1 - (T/T')]ln[1 - L(t)] - [1 + (T/T')]ln \mid L(t) \\ -M^{-2} \mid -(T/T')ln \mid L(t) \mid, \tag{23}$$

where T' is given by

$$T' = \frac{2K_0 T}{p_0(M^2 - 1)}. \tag{24}$$

The ratio T/T' is given by

$$\frac{T}{T'} = \frac{(M^2 - 2)M_v^2}{M^2 - M_v^2}. \tag{25}$$

This equation shows that the range $0 \leq T/T' \leq 1$ corresponds to the range $0 \leq M_v \leq 1$. Figure 1 depicts the variation of $L(t)$ with t/T' for $M = 5$. Figure 2 shows the variation of $L(t)$ with t/T for $1 < T/T' < \infty$ for $M = 5$. Clearly, for $M_v < 1$, there is no shock at all. If the flow is supersonic with respect to the viscoelastic sound speed, a discontinuity in the velocity occurs.

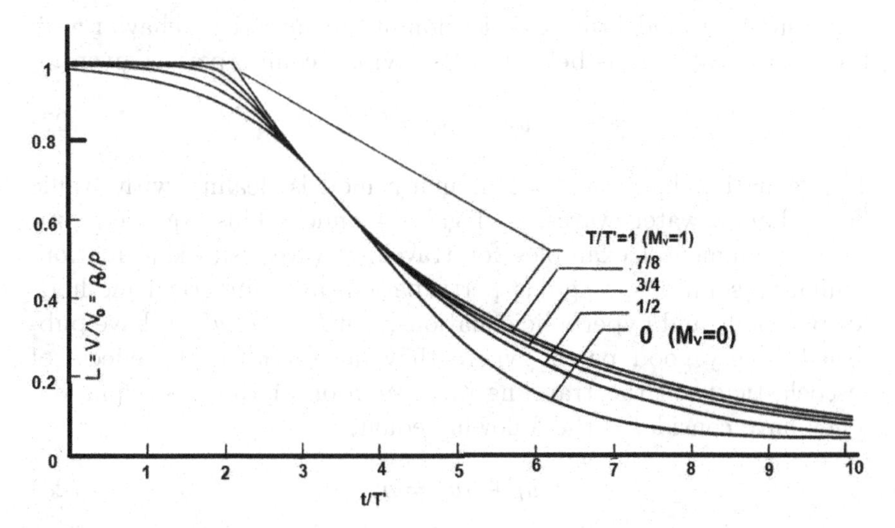

Figure 1. Shock structure for $M_v \leq 1 < M$.

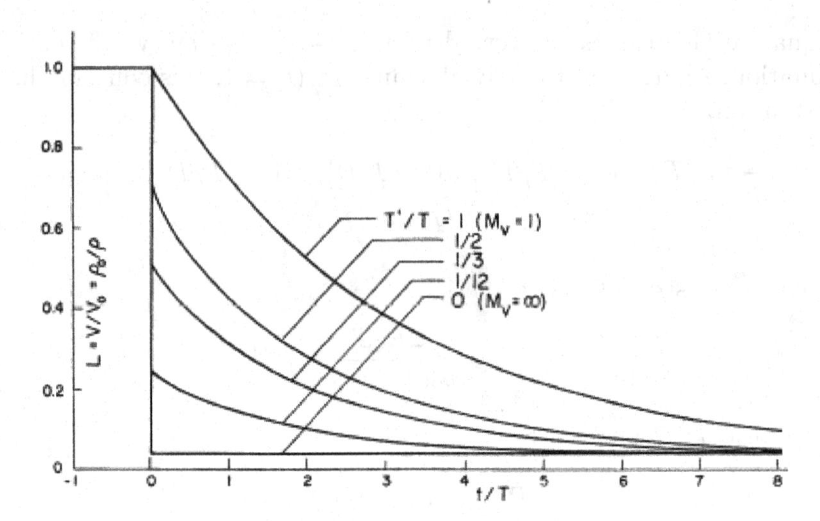

Figure 2. Shock structure for $1 \leq M_v \leq M$.

3. Shock Waves in Viscoelastic Burgers' Equation

The Burgers' equation [1],

$$u_t + uu_x = \epsilon u_{xx}, \tag{26}$$

is the simplest model that couples nonlinear convective behavior with the dissipative viscous behavior. Its inviscid counterpart is given by

$$u_t + uu_x = 0. \tag{27}$$

This equation has appeared in many models dealing with traffic flow, shallow-water waves, and gas dynamics. This equation provides fundamental examples for traveling waves, shock formation, similarity solutions, singular perturbation, and numerical methods for parabolic and hyperbolic equations. Camacho *et al.* [4] have published a very good paper, where they have studied the effects of viscoelasticity on the traveling-wave solution of Burgers' equation. They have considered the following equations:

$$u_t + uu_x = \sigma_x, \tag{28}$$

$$\sigma_t + u\sigma_x - \sigma u_x = \alpha u_x - \beta\sigma. \tag{29}$$

Clearly, the constitutive equation in Eq. (29) resembles the one-dimensional version of the upper-convected Maxwell model [2]. To find the traveling wave solution to the above system of equations, Camacho *et al.* [4] have introduced the new variables $u(x,t) = U(\xi)$, $\sigma(x,t) = S(\xi)$, where $\xi = x - ct$. The above system of equations becomes

$$-cU' + UU' = S', \tag{30}$$

$$-cS' + US' - SU' = \alpha U' - \beta S. \tag{31}$$

The asymptotic boundary conditions are

$$\begin{aligned} U(-\infty) &= u_l, \quad S(-\infty) = 0, \\ U(\infty) &= u_r, \quad S(\infty) = 0. \end{aligned} \tag{32}$$

The authors have given a rigorous existence proof and have reported that a traveling-wave solution exists if and only if

$$u_l > u_r \quad \text{and} \quad \alpha > \left(\frac{u_l - u_r}{2}\right)^2 \tag{33}$$

or

$$u_l < u_r \quad \text{and} \quad 2\alpha < \left(\frac{u_l - u_r}{2}\right)^2. \tag{34}$$

Camacho *et al.* [4] have also solved the aforementioned viscoelastic Burgers' equation by numerical method. Figure 3 shows the wave profiles for $u_l = 2$, $u_r = 0$, $\beta = 1$, and (i) $\alpha = 0.8$ and (ii) $\alpha = 0.25$.

Recently, Ramos [5] has discussed the traveling-wave solution of the following one-dimensional viscoelastic Burgers' equation:

$$u_t + uu_x = \sigma_x, \tag{35}$$

where the stress σ satisfies

$$\sigma + \lambda\frac{\delta\sigma}{\delta t} = \mu u_x, \tag{36}$$

where λ is a constant relaxation time. Then,

$$\frac{\delta\sigma}{\delta t} = \frac{D\sigma}{Dt} - 2au_x\sigma. \tag{37}$$

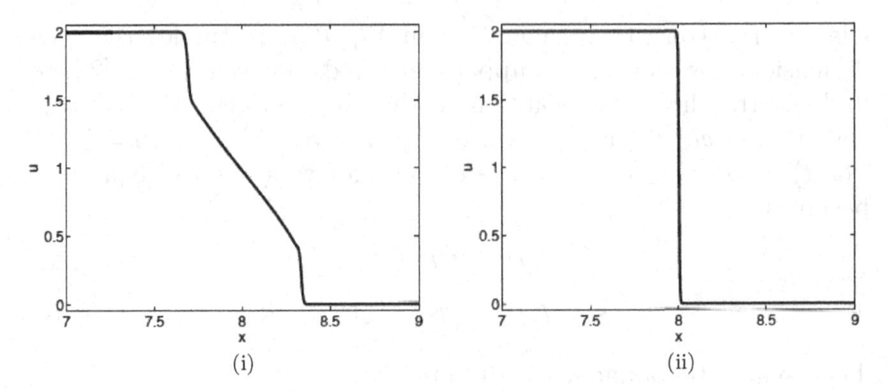

Figure 3. Plots of wave profiles.

The values $a = 1$ and $a = -1$ correspond to lower- and upper-convected Maxwell derivatives. In order to find the traveling-wave solution, one has to substitute $u(t,x) = F(\xi)$ and $\sigma(t,x) = G(\xi)$, where $\xi = x - ct$; c is the wave speed. Now, Eqs. (35)–(37) become

$$G' = (F - c)F', \tag{38}$$

$$G + \lambda((F - c)G' - 2aGF') = \mu F'. \tag{39}$$

Integrating Eq. (38) gives

$$G = \frac{1}{2}F^2 - cF + A \equiv P(F) = \frac{1}{2}(F - c)^2 + A - \frac{1}{2}c^2, \tag{40}$$

where A is the constant of integration. Substituting Eq. (40) in Eq. (39), we obtain

$$P(F) = Q(F)F' \quad Q(F) = \mu + a\lambda(2A - c^2) + \lambda(a - 1)(F - c)^2. \tag{41}$$

Now, denoting $F(-\infty) = F_L$ and $F(\infty) = F_R$ and introducing $f = \frac{F}{F_r}$ and $g = \frac{C}{F_r^2}$, Ramos [5] has carried out a rigorous analysis for the existence of solution. Figures 4–7 show the velocity and stress profiles for different flow parameters.

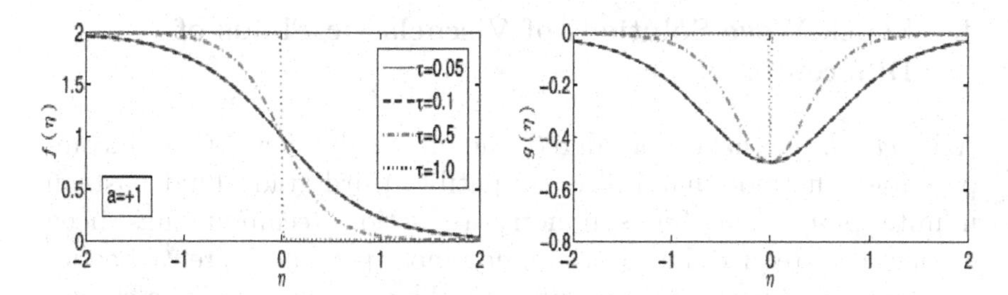

Figure 4. Velocity and stress profiles for $f_L = 2$, $f_R = 0.01$.

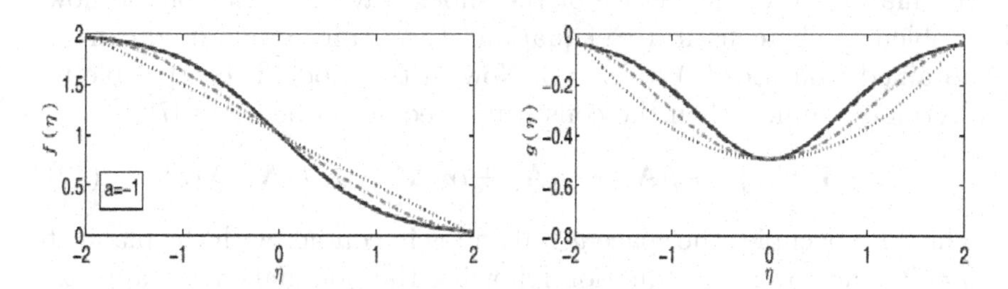

Figure 5. Velocity and stress profiles for $f_L = 2$, $f_R = 0.01$.

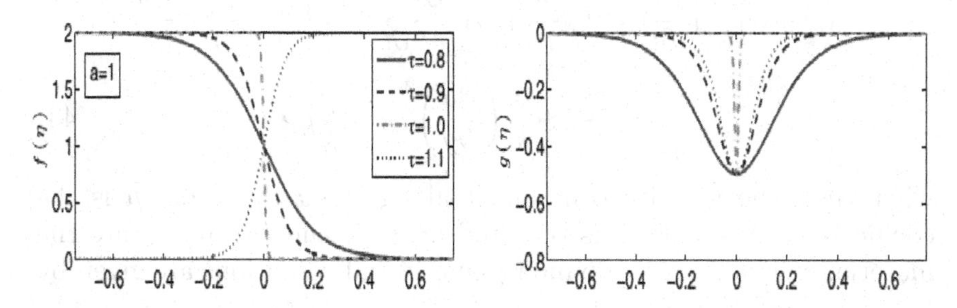

Figure 6. Velocity and stress profiles for $f_L = 2$, $f_R = 0.01$ for different τ.

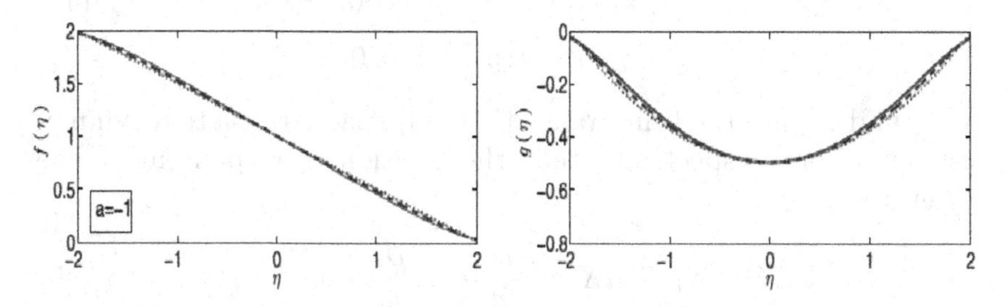

Figure 7. Velocity and stress profiles for $f_L = 2$, $f_R = 0.01$ for different τ.

4. Shock-Wave Solutions of Viscoelastic Fluids of Differential Type

Aziz *et al.* [6] have considered the unsteady flow of an incompressible, thermodynamically compatible third-grade fluid past an infinite plate. The Lie symmetry reduction technique has been employed to reduce the governing nonlinear partial-differential equations into nonlinear ordinary-differential equations. Finally, a closed-form shock-wave solution has been obtained. The results describe the mathematical structure of the shock-wave behavior of the flow problems. The constitutive equation of the third-grade fluid can be obtained from Eq. (3) for $n = 3$. Now, if the model is compatible to thermodynamics, then the constitutive equation becomes [7]

$$\mathbf{T} = -p\mathbf{I} + \mu\mathbf{A}_1 + \alpha_1\mathbf{A}_2 + \alpha_2\mathbf{A}_1^2 + \beta_3(tr\mathbf{A}_1^2)\mathbf{A}_1. \tag{42}$$

The fluid occupies the space $y > 0$ and is in contact with the plane at $y = 0$. The governing equation following the boundary-layer approximations becomes

$$\begin{aligned}\left(\rho + \alpha_1\frac{\varphi}{k}\right)\frac{\partial u}{\partial t} = \mu\frac{\partial^2 u}{\partial y^2} + \alpha_1\frac{\partial^3 u}{\partial y^2 \partial t} + 6\beta_3\left(\frac{\partial u}{\partial y}\right)^2\frac{\partial^2 u}{\partial y^2}\\ -2\beta_3\frac{\varphi}{k}\left(\frac{\partial u}{\partial y}\right)^2 u - \frac{\varphi}{k}\mu u,\end{aligned} \tag{43}$$

where u is the velocity component along the x direction, μ is the coefficient of viscosity, ρ is the fluid density, and α_1 and β_3 are the material constants. The boundary and initial conditions are given by

$$\begin{aligned} u(0,t) &= u_0 g(t), \\ u(\infty,t) &= 0, \qquad t > 0, \\ u(y,0) &= f(y), \quad y > 0. \end{aligned} \tag{44}$$

If X_1 and X_2 are the time translation and space translation symmetry generators, respectively, then the solution corresponding to the generator

$$X = X_1 + mX_2 = \frac{\partial}{\partial t} + m\frac{\partial}{\partial y}, \quad m > 0, \tag{45}$$

would represent traveling-wave solution with constant wave speed m. The characteristic system corresponding to Eq. (45) is

$$\frac{dy}{m} = \frac{dt}{1} = \frac{du}{0}. \tag{46}$$

Solving Eq. (46), the similarity variables are given by

$$u(y,t) = F(\eta), \quad \eta = y - mt. \tag{47}$$

Using these similarity variables in the governing equations in Eq. (43), Aziz *et al.* [6] have found the following nonlinear ordinary-differential equation:

$$\left(\rho + \alpha_1 \frac{\varphi}{k}\right) m \frac{dF}{d\eta} + \mu \frac{d^2F}{d\eta^2} - m\alpha_1 \frac{d^3F}{d\eta^3} + 6\beta_3 \left(\frac{dF}{d\eta}\right)^2 \frac{d^2F}{d\eta^2}$$
$$- 2\beta_3 \frac{\varphi}{k} F \left(\frac{dF}{d\eta}\right)^2 - \mu \frac{\varphi}{k} F = 0. \tag{48}$$

Let us proceed to find the traveling-wave solution of Eq. (47). Assuming the solution of Eq. (47) as $F(\eta) = a\exp(b\eta)$ and following [6], we get the solution of Eq. (47), which is given by

$$F(\eta) = a\exp\left[\sqrt{\frac{\varphi}{3k}}\eta\right]. \tag{49}$$

Hence, the solution of the governing equation in Eq. (43) becomes

$$F(\eta) = a\exp\left[\sqrt{\frac{\varphi}{3k}}(y - mt)\right] \quad m > 0. \tag{50}$$

The solution in Eq. (49) is the shock-wave solution of Eq. (43). Clearly, the solution satisfies all the boundary conditions and shows the shock-wave behavior of the flow problem. The velocity gradient $\frac{\partial u}{\partial y} \to \infty$ as $y \to \infty$.

The second problem considered by Aziz *et al.* [6] is the unsteady Magnetohydrodynamics (MHD) flow of third-grade fluid in a porous medium. The boundary-layer equation is given by

$$\left(\rho+\alpha_1\frac{\varphi}{k}\right)\frac{\partial u}{\partial t}=\mu\frac{\partial^2 u}{\partial y^2}+\alpha_1\frac{\partial^3 u}{\partial y^2\partial t}+6\beta_3\left(\frac{\partial u}{\partial y}\right)^2\frac{\partial^2 u}{\partial y^2}$$
$$-2\beta_3\frac{\varphi}{k}\left(\frac{\partial u}{\partial y}\right)^2 u-\frac{\varphi}{k}\mu u-\sigma B_0^2 u, \qquad (51)$$

where σ is the electrical conductivity and B_0 is the applied uniform magnetic field. The above equation has to be solved subject to no-slip boundary conditions similar to Eq. (44). Using the Lie symmetry analysis as above, one obtains the following similarity equation:

$$\left(\rho+\alpha_1\frac{\varphi}{k}\right)m\frac{dF}{d\eta}+\mu\frac{d^2F}{d\eta^2}-m\alpha_1\frac{d^3F}{d\eta^3}+6\beta_3\left(\frac{dF}{d\eta}\right)^2\frac{d^2F}{d\eta^2}$$
$$-2\beta_3\frac{\varphi}{k}F\left(\frac{dF}{d\eta}\right)^2-\mu\frac{\varphi}{k}F-\sigma B_0^2F=0. \qquad (52)$$

Equation (51) admits the shock wave solution $F(\eta)=a\exp\left[\sqrt{\frac{\varphi}{3k}}\eta\right]$ (same as Eq. (49)). Hence, the solution of the governing equation in Eq. (50) is given by

$$F(\eta)=a\exp\left[\sqrt{\frac{\varphi}{3k}}(y-mt)\right], \quad m>0. \qquad (53)$$

The third related problem considered by Aziz *et al.* [6] is the unsteady MHD flow of a third-grade fluid past a porous plate with suction/injection velocity. They have obtained a solution for the resulting boundary-layer equation similar to Eq. (52). It is evident that the shock-wave solution of the three governing equations are same, but the imposing conditions on the physical parameters are different. This means, in each case, the speed of the traveling shock wave is different. Figures 8 and 9 show the two-dimensional and three-dimensional profiles of the shock-wave solution for $\varphi=4,\ k=0.2$, $m=1$, and $t=\pi/2$.

Aziz *et al.* [8] have also studied the shock-wave solution of viscoelastic fourth-grade fluid past an infinite plate. The constitutive equation of the fourth-grade fluid is given by Eq. (3). Sahoo and Poncet [9] have obtained the numerical solution for a similar kind of flow problem. The plate coincides with x axis, and the y axis is

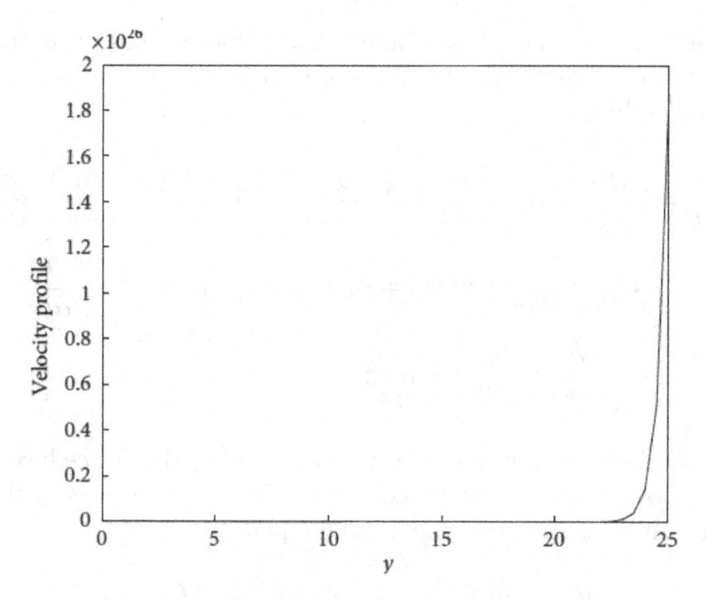

Figure 8. Profile of the shock-wave solutions.

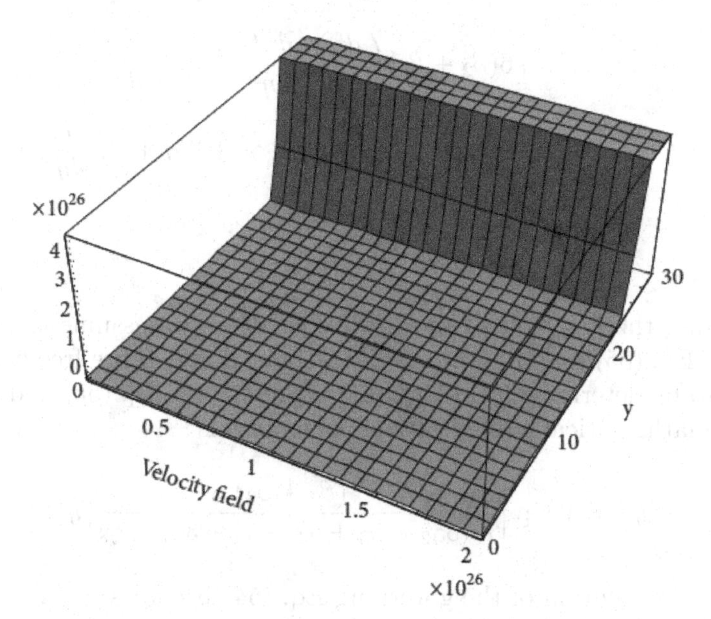

Figure 9. 3-D profile of the shock-wave solutions.

perpendicular to the plate. Taking the velocity field $(u(y,t),0,0)$, the equation of continuity is identically satisfied and the momentum equation reduces to

$$\rho\frac{\partial u}{\partial t}=\mu\frac{\partial^2 u}{\partial y^2}+\alpha_1\frac{\partial^3 u}{\partial y^2\partial t}+\beta_1\frac{\partial^4 u}{\partial y^2\partial t^2}+6(\beta_2+\beta_3)\left(\frac{\partial u}{\partial y}\right)^2\frac{\partial^2 u}{\partial y^2}$$
$$+\gamma_1\frac{\partial^5 u}{\partial y^2\partial t^3}+2(3\gamma_2+\gamma_3+\gamma_4+\gamma_5+3\gamma_7+\gamma_8)\frac{\partial}{\partial y}$$
$$\times\left[\left(\frac{\partial u}{\partial y}\right)\frac{\partial^2 u}{\partial y\partial t}\right]-\sigma B_0^2 u. \tag{54}$$

Using the Lie symmetry method and following the procedure mentioned above, the momentum equation (Eq. (53)) reduces to the following nonlinear ordinary-differential equation:

$$-m\rho\frac{df}{d\eta}=\mu\frac{d^2 f}{d\eta^2}-\alpha_1 m\frac{d^3 f}{d\eta^3}+\beta_1 m^2\frac{d^4 f}{d\eta^4}$$
$$+6(\beta_2+\beta_3)\left(\frac{df}{d\eta}\right)^2\frac{d^2 f}{d\eta^2}-\gamma_1 m^3\frac{d^5 f}{d\eta^5}$$
$$-2m(3\gamma_2+\gamma_3+\gamma_4+\gamma_5+3\gamma_7+\gamma_8)\frac{d}{d\eta}$$
$$\times\left[\left(\frac{df}{d\eta}\right)^2\frac{d^2 f}{\eta^2}\right]. \tag{55}$$

Following the principle of shock-wave solution, we assume the solution of Eq. (55) as $f(\eta)=A\,exp(B\eta)$. Here, A and B are free parameters to be determined. Putting this expression in Eq. (55) and after some mathematical simplifications, one obtains

$$f(\eta)=A\exp\left[\frac{(\beta_2+\beta_3)}{m(3\gamma_2+\gamma_3+\gamma_4+\gamma_5+3\gamma_7+\gamma_8)}\eta\right]. \tag{56}$$

Hence, the solution of the governing Eq. (54) becomes

$$f(\eta)=A\exp\left[\frac{(\beta_2+\beta_3)(y-mt)}{m(3\gamma_2+\gamma_3+\gamma_4+\gamma_5+3\gamma_7+\gamma_8)}\right],\quad m>0. \tag{57}$$

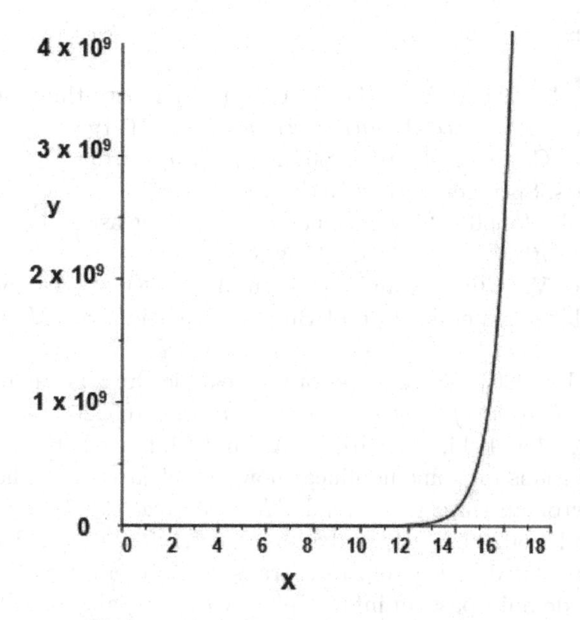

Figure 10. 2-D shock-wave solution.

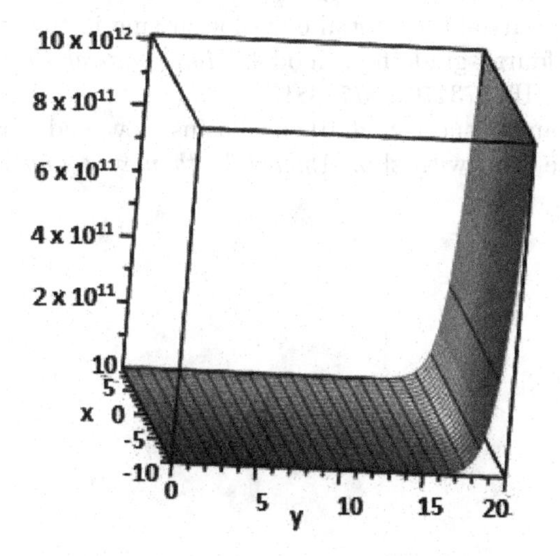

Figure 11. 3-D shock-wave solution.

Figures 10 and 11 show the shock-wave behavior of solution (57). This solution shows the hidden shock-wave behavior with $\frac{\partial u}{\partial y} \to \infty$ as $y \to \infty$.

References

[1] Burgers, J. (1948). A mathematical model illustrating the theory of turbulence. *Advanced Applied Mechanics*, 1, 171–199.

[2] Truesdell, C. and Noll, W. (2004). *The Non-Linear Field Theories of Mechanics*, Springer-Verlag: Berlin Heidelberg.

[3] Pipkin, A. (1966). Shock structure in viscoelastic fluid. *Quarterly Applied Mathematics*, 23(4), 297–303.

[4] Camacho, V., Guy, R., and Jacobsen, J. (2008). Waves and shocks in a viscoelastic generalization of Burgers' equation. *SIAM*, 68(5), 1316–1332.

[5] Ramos, J. (2020). Shock waves of viscoelastic Burgers equations. *International Journal of Engineering Science*, 149, 103226.

[6] Aziz, T., Moitsheki, R., Fatima, A., and Mahomed, F. (2013). Shock wave solutions for some nonlinear flow models arising in the study of a non-Newtonian third grade fluid. *Mathematical Problems in Engineering* (Art ID 602902). http://dx.doi.org/10.1155/2013/602902

[7] Sahoo, B. (2010). Flow and heat transfer of an electrically conducting third grade fluid past an infinite plate with partial slip. *Meccanica*, 45, 319–330.

[8] Aziz, T., Fatima, A., and Mahomed, F. (2013). Shock wave solution of nonlinear partial differential equation arising in the study of a non-Newtonian fourth grade fluid model. *Mathematical Problems in Engineering* (Art ID 573170), 877–881.

[9] Sahoo, B. and Poncet S. (2013). Blasius flow and heat transfer of fourth-grade fluid with slip. *Applied Mathematics and Mechanics*, 34, 1465–1480.

https://doi.org/10.1142/9789811245367_0008

Chapter 8

Wave Propagation through Resonators, Resonators in Series and Multi-Resonator

J. Dandsena* and D. P. Jena†

Department of Industrial Design,
National Institute of Technology,
Rourkela, India
* *Janmenjay888@gmail.com*
† *jena.dibya@gmail.com*

Abstract

One-dimensional acoustic wave propagation through resonators of sub-wavelength size has its own importance because of novel physical phenomenon. Single resonators have a very narrow operational band, and multiple resonators are suggested for broadband operation. We present an investigation on wave propagation through a single resonator, periodically arranged resonators, and a multi-resonator unit structure. To achieve wide operational band, a periodic arrangement of resonators connected in series and a unique arrangement of multiple resonators as a single unit are proposed. For the proposed models, transmission loss and effective metamaterial properties are analyzed through reflected and transmitted data. Effective bulk modulus and effective density are considered to calculate and study the metamaterial behavior. All the investigations are performed through analytical, numerical, and experimental process. Transmission loss tube is used to perform all the experiments, and the resonators are mounted over a waveguide. ASTM E-2611 is followed to calculate the transfer matrix, and the subsequent calculations are conducted from that data. Existence of resonance plays a rich role and provides fundamentally different modes for the band structure. The results clearly show that the presence of a resonator or multiple resonators and their arrangements can play a prominent role in the design any acoustic band-gap materials.

Keywords: Transfer matrix, scattering matrix, resonator, resonance frequency, multi resonator

1. Introduction

Resonators are frequently used to attenuate noise propagation in a narrow frequency band. Helmholtz Resonator (HR) is a familiar acoustic model which connects itself to a wave guide and show high transmission loss over its resonance frequency. Extensive research has been carried out to improve the sound-attenuation performance. Many achievements have been accomplished and are registered in numerous literature.

The performance of the HR has been enhanced by proper design and modification of geometrical parameters. The extended neck and spiral neck designs are proposed by Cai *et al.* to handle space constraints [1]. Dual HRs, where two HRs are connected in series, are analyzed [2]. Similarly, periodic arrangement of HRs are also inspected for hybrid noise attenuation [3]. Sound transmission loss has been used as a key parameter index, and HRs are evaluated for their performance [4, 5].

HRs are also studied for their effective metamaterial properties, such as effective bulk modulus and effective density [6, 7]. Homogenization approach has been adopted to develop analytical models and calculate effective properties [8]. Metamaterial effective properties have also been calculated by correlating electromagnetic models to acoustic analogy and adopting acoustic transmission line approach [9].

This chapter explains the acoustic wave characteristics inside a HR during a one-dimensional flow. The behavior of the resonator and its working frequency are examined. An array of such HRs is arranged in series and its characteristics are explored. A special multi-resonator system is proposed in search of a wider working frequency range. Finally, the effective metamaterial properties of all the designs are investigated.

2. Design of Resonators

The performance of a single resonator is directly influenced by its resonance frequency and is limited to a single frequency. A HR coupled with a duct, as shown in Figure 1(a), is considered for investigation in this research. This acoustic system can be represented as a mechanical lumped-parameter model, as shown in Figure 1(b). The incident acoustic wave inside the duct pushes the air mass of the neck into the cavity. This increases the pressure of the cavity, and the air mass is again pushed back. Due to this inertia, the air mass is lifted from its static equilibrium position. This leaves low pressure inside the cavity, and the incident pressure pushes the air mass back into the cavity again. Thus, the air present in the cavity acts like a spring during this oscillation process. Damping loss occurs due to the relative motion of the air mass against the internal surface of the neck.

The resonance frequency of a HR is determined by $f_{HR} = \frac{c_0}{2\pi}\sqrt{\frac{A_n}{V_c l'_n}}$, where c_0 is the speed of sound in the medium (air), A_n is the cross-sectional area of the neck, l'_n is the effective length of the neck, and V_c is the volume of the cavity. This relation clearly tells that the resonance frequency is only determined by the geometry of the HR. Hence, it is easy to design a HR with desired resonance frequency. The acoustic impedance can be obtained through spring–mass lumped-parameter model and is given as

$$Z_{HR} = \frac{j\rho_0 l'_n (f_{HR}^2 - \omega^2)}{A_n f_{HR}}, \tag{1}$$

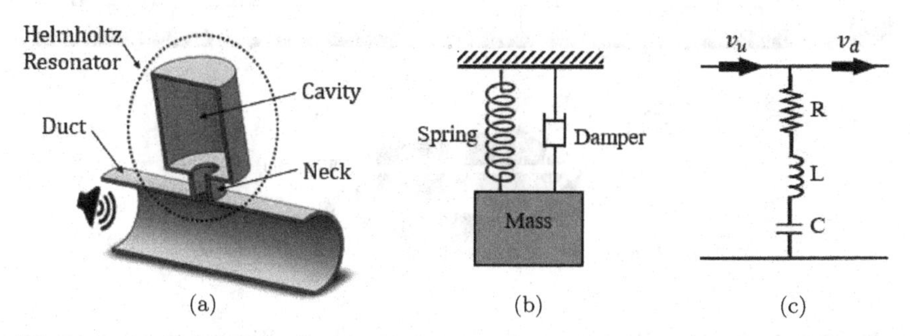

Figure 1. Helmholtz resonator and its mechanical and electrical lumped-element models.

where ρ_0 is the density of the medium (air) and ω is the angular frequency.

A HR can also be represented as an electrical lumped-parameter model, as shown in Figure 1(c). The velocity of a one-dimensional acoustic wave can be matched up with the current flow in an electrical transmission line and the acoustic pressure difference as the voltage difference across electrical terminals. The HR cavity acts as a capacitor, with capacitance $C = V_c/(\rho_0 c_0^2)$, and the neck acts as an inductor, with inductance $L = (\rho_0 l_{ne})/A_n$, where ρ_0 is the density of the fluid medium, and the resistance offered by the neck is $R = \rho_0 c_0/A_n$. Analyzing this electrical lumped-element model, the impedance can be calculated as

$$Z_{HR} = \frac{\omega^2}{\pi c_0} + j\omega \frac{l_{ne}}{A_n} + \frac{c_0^2}{j\omega V_c} \tag{2}$$

To improve the bandwidth of working frequency, an array of such HRs is prepared by arranging them in series, as shown in Figure 2(a). It is fabricated with 10 HRs of similar size attached to a cylindrical waveguide of 45 mm diameter. The wave guide and the resonators are made up of aluminum. The cavity volume is $9.5426 \times 10^4\,\text{mm}^3$, the neck cross-sectional area and length are $177\,\text{mm}^2$ and 15.5 mm, respectively.

A unique multi-resonator unit cell is designed within a confined space to achieve effective frequency bands at various frequencies. Three cavities with narrow openings (necks) are combined together,

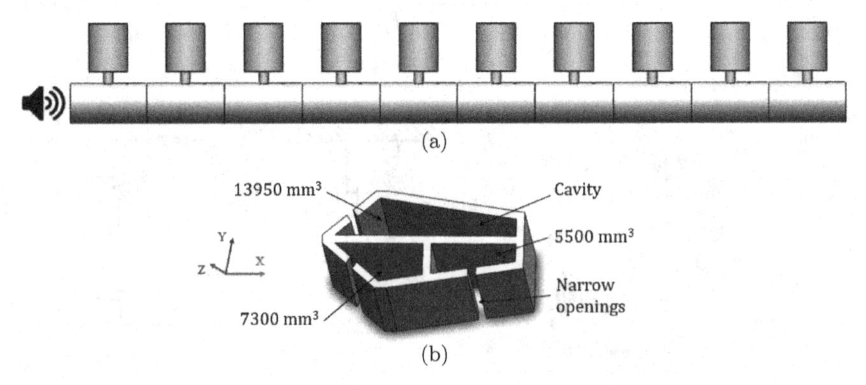

Figure 2. (a) Ten HRs arranged in series and (b) multi-resonator unit cell.

Figure 3. (a) HR, (b) 10 HRs arranged in series, and (c) multi-resonator unit cell.

as shown in Figure 2(b). The sample is fabricated with a three-dimensional printing machine (Dimension SST 1200es, Stratasys), and the material used is Acrylonitrile Butadiene Styrene (ABS) as shown in Figure 3.

3. Performance of Resonators

3.1. *Transmission Loss*

Transfer matrix (TM) modeling is adopted to investigate the transmission loss of the designed samples as shown in Figure 4. The TM of a HR mounted on a one-dimensional wave guide with uniform cross-section and rigid walls can be calculated as follows:

$$\begin{bmatrix} T_{11} & T_{12} \\ T_{22} & T_{21} \end{bmatrix} = [T_{\text{duct}}][T_{HR}][T_{\text{duct}}], \tag{3}$$

where T_{duct} is the TM of the duct of length l and cross-sectional area A_d, and T_{HR} is the TM of the HR, given by

$$[T_{\text{duct}}] = \begin{bmatrix} \cos(k_0 l) & jc_0\sin(k_0 l)/A_d \\ jA_d\sin(k_0 l)/c_0 & \cos(k_0 l) \end{bmatrix}, \tag{4}$$

$$[T_{HR}] = \begin{bmatrix} 1 & 0 \\ 1/\left(\frac{\omega^2}{\pi c_0} + \frac{j\omega l'_n}{A_n} + \frac{c_0^2}{j\omega V_c}\right) & 1 \end{bmatrix}. \tag{5}$$

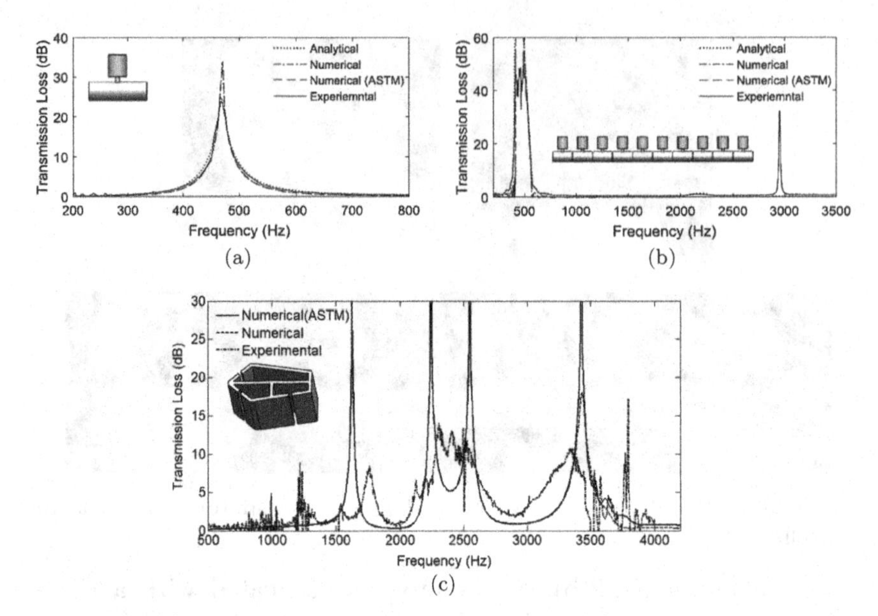

Figure 4. (a) Transmission loss of single resonator, (b) array of resonators, and (c) multiresonator.

From the TM, the transmission loss can be calculated as follows:

$$TL = 20\log_{10}\left|\frac{T_{11} + T_{12}/(\rho_0 c_0) + T_{21}\rho_0 c_0 + T_{22}}{2e^{jkd}}\right|. \tag{6}$$

3.2. *Effective Metamaterial Properties*

Acoustic resonators are always said to exhibit metamaterial properties [7]. To investigate the effective metamaterial properties of the designed samples, the following analytical equations are used. The investigation starts with the calculation of scattering parameters which hold the information about reflected and transmitted waves. Assuming the samples as homogeneous materials, the scattering parameters can be calculated from TM data [10].

$$S_{11} = S_{22} = \frac{T_{11} + \left(\frac{T_{12}}{Z_0}\right) - T_{21}Z_0 - T_{22}}{T_{11} + \left(\frac{T_{12}}{Z_0}\right) + T_{21}Z_0 + T_{22}}, \tag{7}$$

$$S_{12} = S_{21} = \frac{2}{T_{11} + \left(\frac{T_{12}}{Z_0}\right) + T_{21}Z_0 + T_{22}}. \tag{8}$$

The impedance (Z_e) and refractive index (n) can be determined from the above calculated parameters [11]:

$$Z_e = \pm\sqrt{\frac{(1+S_{11})^2 - S_{21}^2}{(1-S_{11})^2 - S_{21}^2}}, \tag{9}$$

$$n = \pm\frac{1}{kd}\cos^{-1}\left(\frac{1}{2S_{21}}\left(1 - S_{11}^2 + S_{21}^2\right) + 2\pi m\right), \tag{10}$$

where m is an integer indicating the branch number.

The effective density and effective bulk modulus of the acoustic samples can be obtained as follows:

$$\rho_e = nZ_{\text{eff}}, \tag{11}$$

$$K_e = n/Z_{\text{eff}}. \tag{12}$$

To validate the method adopted, Finite-Element Model (FEM) is implemented through commercially available COMSOL Multiphysics© software. To simulate the domain, the models are prepared in a 3-D environment, and the physics-controlled mesh is applied, with the maximum element size kept $\lambda/15$ of the maximum investigated frequency. Plane-wave radiation is applied at one end and the other end is applied with impedance boundary condition. The outer surfaces of the model are set as sound-hard boundary wall. The contours of the simulated samples are shown in Figure 5.

ASTM E-2611 is followed to obtain TM from the FEM simulation. Boundary probes are used to mimic the four microphones as described in the ASTM E-2611 as shown in the Figure 6. Sound-hard wall and impedance boundary physics are applied at the outlet end to imitate blocked and anechoic states, respectively. Succeeding the TM calculation, scattering parameters is acquired using Eqs. (7) and (8). The effective density and effective bulk-modulus is then calculated through Eqs. (9)–(12).

Experiments are also conducted to make sure the physical existence of metamaterials and their behavior. Transmission loss tube, based on ASTM E-2611, is used to conduct all the experiments. The transmission loss tube consists four microphones and support two

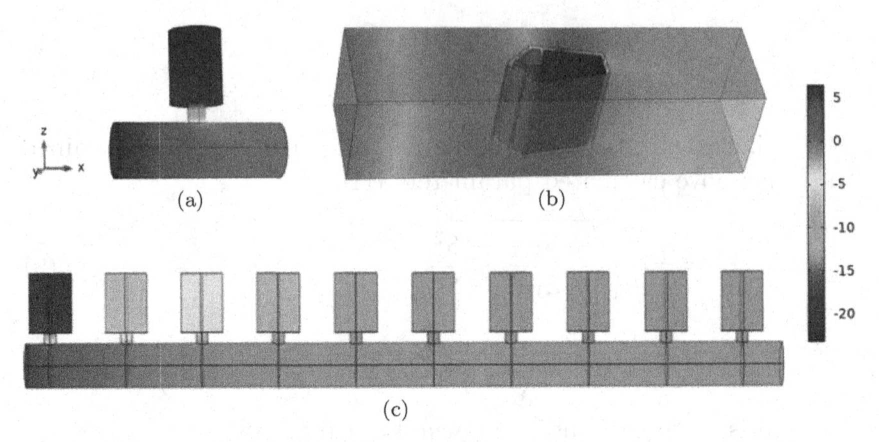

Figure 5. Contours of the three samples.

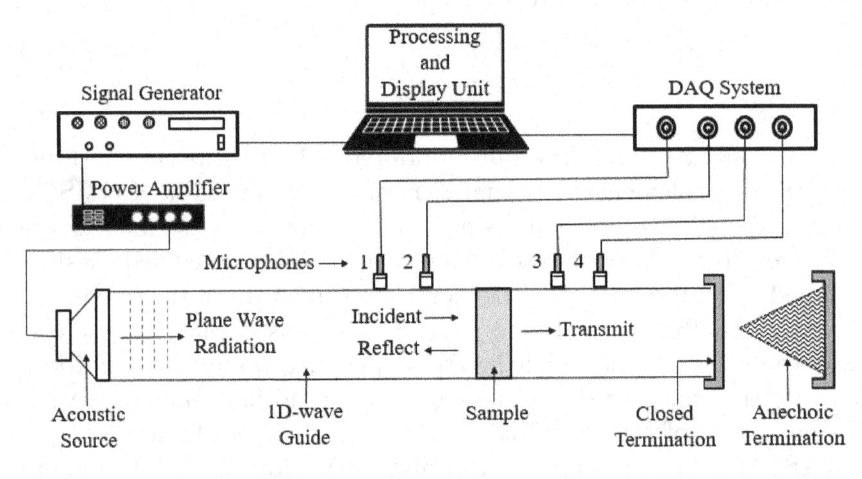

Figure 6. Schematic of the test setup as per ASTM E-2611.

load conditions. Ahuja© made noise source, PCB© made microphones, and NI© made data acquisition system are installed in the experiment setup. LabVIEW© software package, installed in the processing and display unit, is used to handle and process the acquired pressure data through the microphones. The blocked setup is obtained by simply blocking the termination end, and the anechoic termination is obtained by simply keeping the termination end open. The prepared samples, resonator, and resonators in series are tested through a transmission loss tube of circular cross-sectional area. The test setups are shown in Figure 7(a) and (b). The multi-resonator unit

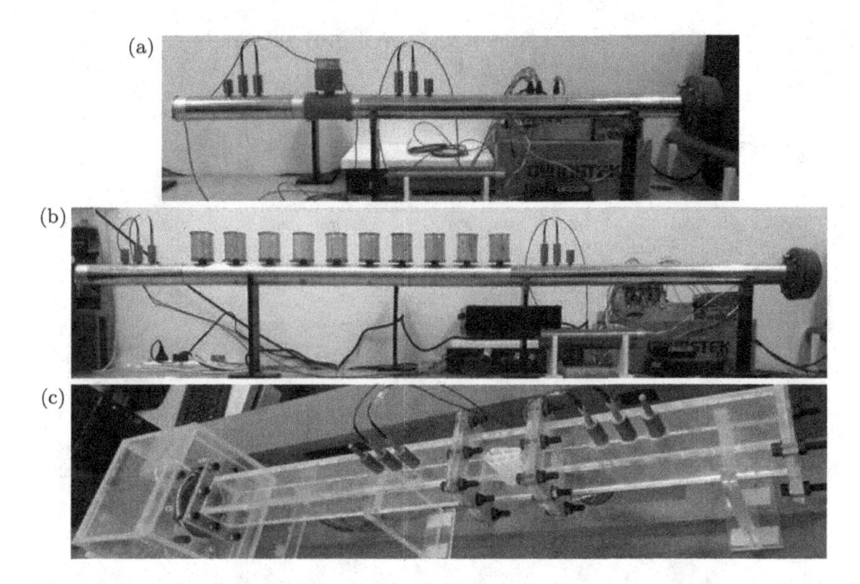

Figure 7. Experimental test setup (circular and square impedance tubes).

cell is tested inside a transmission loss tube of square cross-sectional area as presented in Figure 7(c). The TM can be accurately obtained from this test facility. Equations (7)–(12) can be used to extract the effective metamaterial properties. The effective bulk modulus and the effective density as a parameter of metamaterial are shown in Figure 8.

4. Conclusion

The transmission loss and effective metamaterial parameters retrieval is accomplished through analytical, numerical, and experimental methods. The analytical equations are discussed with proper technique and steps followed by finite-element model simulation. The adopted simulation technique is evaluated, and the results fit well with the analytical solution. Furthermore, the experimental outcome adds an additional assurance over the analytical and numerical methods.

From the transmission loss analysis, it can be observed that the HR has a single peak at its resonance frequency. Whereas, in the case of array of HRs arranged in series, the bandwidth of the transmission loss curve increases, which arises from the equivalent losses of the

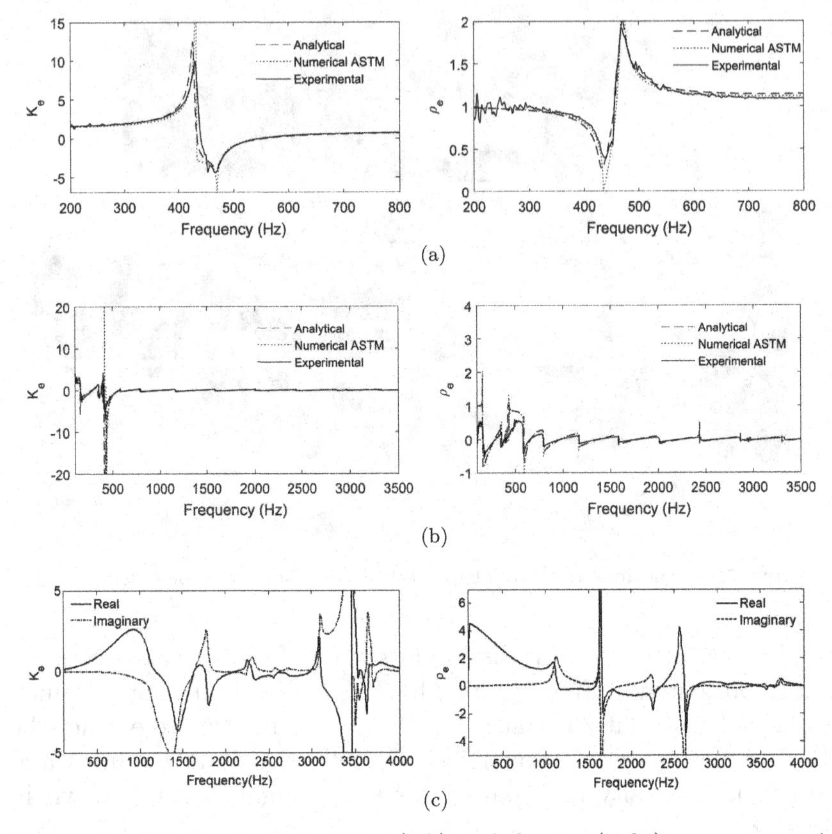

Figure 8. Effective bulk modulus (left) and density (right) as metamaterial properties of (a) single resonator, (b) array of resonator, and (c) multi-resonator.

HRs. In the case of multi-resonator unit, the transmission loss has four resonator peaks. The multi-resonator has three cavities exhibiting three transmission loss peaks, and the extra resonance peak is the result of the unit cell shape inside the square duct.

Effective density and effective bulk modulus are extracted from the designed samples. The HR comes out as a single negative metamaterial as its effective bulk modulus is negative. The negative characteristic is observed at the resonance frequency. The array of HRs comes out as double negative as the system shows negative values for effective bulk modulus and density. Effective bulk modulus is negative because the acoustic wave response is in out-of-phase oscillation with respect to the applied pressure. The multi-resonator unit is also observed as a double-negative metamaterial.

References

[1] Cai, C., Mak, C. M., and Shi, X. (2017). An extended neck versus a spiral neck of the Helmholtz resonator. *Applied Acoustics*, 115, 74–80. doi: 10.1016/j.apacoust.2016.08.020.
[2] Xu, M. B., Selamet, A., and Kim, H. (2010). Dual Helmholtz resonator. *Applied Acoustics*, 71(9), 822–829. doi: 10.1016/j.apacoust.2010.04.007.
[3] Cai, C. and Mak, C. M. (2018). Hybrid noise control in a duct using a periodic dual Helmholtz resonator array. *Applied Acoustics*, 134(December 2017), 119–124. doi: 10.1016/j.apacoust.2018.01.015.
[4] Cai, C. and Mak, C. M. (2018). Noise attenuation capacity of a Helmholtz resonator. *Advances in Engineering Software*, 116(October 2017), 60–66. doi: 10.1016/j.advengsoft.2017.12.003.
[5] Li, L., Liu, Y., Zhang, F., and Sun, Z. (2017). Several explanations on the theoretical formula of Helmholtz resonator. *Advances in Engineering Software*, 114, 361–371. doi: 10.1016/j.advengsoft.2017.08.004.
[6] Nemati, N., Kumar, A., Lafarge, D., and Fang, N. X. (2015). Nonlocal description of sound propagation through an array of Helmholtz resonators. *Comptes Rendus Mécanique*, 343(12), 656–669. doi: 10.1016/j.crme.2015.05.001.
[7] Fang, N. *et al.*, (2006). Ultrasonic metamaterials with negative modulus. *Nature Materials*, 5(6), 452–456. doi: 10.1038/nmat1644.
[8] Hu, X., Ho, K. M., Chan, C. T., and Zi, J. (2008). Homogenization of acoustic metamaterials of Helmholtz resonators in fluid. *Physical Review B: Condensed Matter and Materials Physics*, 77(17), 2–5. doi: 10.1103/PhysRevB.77.172301.
[9] Cheng, Y., Xu, J. Y., and Liu, X. J. (2008). One-dimensional structured ultrasonic metamaterials with simultaneously negative dynamic density and modulus. *Physical Review B: Condensed Matter and Materials Physics*, 77(4), 1–10. doi: 10.1103/PhysRevB.77.045134.
[10] Seo, S.-H. and Kim, Y.-H. (2005). Silencer design by using array resonators for low-frequency band noise reduction. *Journal of the Acoustical Society of America*, 118(4), 2332–2338. doi: 10.1121/1.2036222.
[11] Szabó, Z., Park, G. H., Hedge, R., and Li, E. P. (2010). A unique extraction of metamaterial parameters based on Kramers-Kronig relationship. *IEEE Transactions on Microwave Theory and Techniques*, 58(10), 2646–2653. doi: 10.1109/TMTT.2010.2065310.

https://doi.org/10.1142/9789811245367_0009

Chapter 9

Solution of Interval-Modified Kawahara Differential Equations using Homotopy Perturbation Transform Method

P. Karunakar*,‡ and S. Chakraverty†,§

**Department of Mathematics, Anurag University, Hyderabad, India*

†Department of Mathematics, National Institute of Technology, Rourkela, Odisha, India

‡karunakarperumandla@gmail.com

§sne_chak@yahoo.com

Abstract

This chapter provides the solution of interval-modified Kawahara Differential Equation (mKDE) that describes the nonlinear water waves in long-wavelength regime. Homotopy perturbation transform method (HPTM) has been used to handle interval-differential equations. The results obtained are compared graphically with crisp results and existing results in the literature.

Keywords: Modified Kawahara equation, homotopy perturbation transform method, interval-differential equations, interval solution

1. Introduction

In the past few decades, nonlinear equations and their solutions has grabbed the attention of researchers. These nonlinear equations act as governing equations of many physical phenomena, such as convection, diffusion, and dispersion. Kawahara differential equations are of that kind and are extensively used for modeling one-dimensional (1-D) propagation of long waves with small amplitude, magneto-acoustic waves in plasma, capillary–gravity waves, and shallow water waves. In this regard, a good number of contributions may be found using various methods. As such, Sirendaoreji [1] used tanh-function method for finding new exact traveling wave solutions of Kawahara equation and mKDE. Modified tanh-coth method has been used by Jabbari and Kheiri [2] for establishing new exact traveling wave solutions of Kawahara equation and mKDE. Wazwaz [3] derived soliton solutions by treating the mKDE using sine–cosine method, tanh method, and extended-tanh method. Kuruly [4] obtained approximate solution of mKDE using homotopy analysis method. Family of traveling wave solution of Kawahara equation has been obtained by homotopy analysis method in [5]. Optimal homotopy analysis method has been applied by Kashkari [6] and Wang [7] for finding a solution of mKDE.

The modified Kawahara equation [4, 8] reads as below:

$$u_t + u^2 u_x + a u_{xxx} + b u_{xxxxx} = 0. \tag{1}$$

2. Homotopy Perturbation Transform Method

In this section, Homotopy Perturbation Transform Method (HPTM) is described briefly. HPTM is the combination of Laplace transform method and homotopy perturbation method, which can be applied for solving Nonlinear Partial-Differential Equations (NPDEs).

Let us consider a general NPDE with the source term $g(x,t)$ to illustrate the basic idea of HPTM as below [9–12]:

$$Du(x,t) + Ru(x,t) + Nu(x,t) = g(x,t), \tag{2}$$

subject to initial conditions

$$u(x,0) = h(x), \quad \frac{\partial}{\partial t} u(x,0) = f(x), \tag{3}$$

where D is the linear differential operator: $D = \frac{\partial^2}{\partial t^2}$ (or $\frac{\partial}{\partial t}$), R is the linear differential operator whose order is less than that of D, and N is the nonlinear differential operator.

The HPTM methodology consists of mainly two steps. The first step is applying Laplace transform on both sides of Eq. (2), and the second step is applying homotopy perturbation method, where decomposition of the nonlinear term is done using He's polynomials.

First, by operating Laplace transform on both sides of (2), we obtain

$$\mathcal{L}[Du(x,t)] = -\mathcal{L}[Ru(x,t)] - \mathcal{L}[Nu(x,t)] + \mathcal{L}[g(x,t)].$$

Assuming that D is a second-order differential operator and using differentiation property of Laplace transform, we get

$$\begin{aligned} s^2\mathcal{L}[u(x,t)] - s\,h(x) - f(x) &= -\mathcal{L}[Ru(x,t)] - \mathcal{L}[Nu(x,t)] \\ &\quad + \mathcal{L}[g(x,t)] \\ \mathcal{L}[u(x,t)] &= \frac{h(x)}{s} + \frac{f(x)}{s^2} + \frac{1}{s^2}\mathcal{L}[g(x,t)] \\ &\quad - \frac{1}{s^2}\mathcal{L}[Ru(x,t)] - \frac{1}{2}\mathcal{L}[Nu(x,t)]. \end{aligned} \tag{4}$$

Applying inverse Laplace transform on both sides of (4), we have

$$u(x,t) = G(x,t) - \mathcal{L}^{-1}\left[\frac{1}{s^2}\mathcal{L}[Ru(x,t)] + \frac{1}{s^2}\mathcal{L}[Nu(x,t)]\right], \tag{5}$$

where $G(x,t)$ is the term arising from the first three terms of the right-hand side of (4).

Next, to apply HPM, first we need to assume solution as a series that contains embedding parameter $p \in [0,\ 1]$ as follows:

$$u(x,t) = \sum_{n=0}^{\infty} p^n u_n(x,t), \tag{6}$$

and the nonlinear term may be decomposed using He's polynomials as

$$Nu(x,t) = \sum_{n=0}^{\infty} p^n H_n(u), \tag{7}$$

where $H_n(u)$ represents the He's polynomials [18–22], which are defined as follows:

$$H_n(u_0, u_1, \ldots, u_n) = \frac{1}{n!}\left[\frac{\partial^n}{\partial p^n} N\left(\sum_{i=0}^{\infty} p^i u_i\right)\right]_{p=0}, \quad n = 0, 1, 2, 3, \ldots. \tag{8}$$

Substituting Eqs. (6) and (7) in Eq. (5) and combining Laplace transform with HPM, one may obtain the following expression:

$$\sum_{n=0}^{\infty} p^n u_n(x,t) = G(x,t) - p\left(\mathcal{L}^{-1}\left[\frac{1}{s^2}\mathcal{L}\left[R\sum_{n=0}^{\infty} p^n u_n(x,t)\right] + \frac{1}{s^2}\mathcal{L}\left[\sum_{n=0}^{\infty} p^n H_n(u)\right]\right]\right). \tag{9}$$

Comparing the coefficients of like powers of "p" on both sides of (9), we may obtain the following successive approximations:

$$p^0 : u_0(x,t) = G(x,t),$$

$$p^1 : u_1(x,t) = \mathcal{L}^{-1}\left[\frac{1}{s^2}\mathcal{L}[Ru_0(x,t)] + \frac{1}{s^2}\mathcal{L}[H_0(u)]\right],$$

$$p^2 : u_2(x,t) = \mathcal{L}^{-1}\left[\frac{1}{s^2}\mathcal{L}[Ru_1(x,t)] + \frac{1}{s^2}\mathcal{L}[H_1(u)]\right],$$

$$p^3 : u_3(x,t) = \mathcal{L}^{-1}\left[\frac{1}{s^2}\mathcal{L}[Ru_2(x,t)] + \frac{1}{s^2}\mathcal{L}[H_2(u)]\right],$$

$$\vdots$$

$$p^n : u_n(x,t) = \mathcal{L}^{-1}\left[\frac{1}{s^2}\mathcal{L}[Ru_{n-1}(x,t)] + \frac{1}{s^2}\mathcal{L}[H_{n-1}(u)]\right],$$

$$\vdots$$

Finally, the solution of the differential equation (2) may be obtained as follows:

$$u(x,t) = \lim_{p\to 1} u_n(x,t) = u_0(x,t) + u_1(x,t) + u_2(x,t) + \cdots. \tag{10}$$

3. Solution of Interval-Modified Kawahara Equation

Here, we apply HPTM method to modified Kawahara differential equation (1) with the initial condition as [1, 13]

$$u(x,0) = \frac{3a}{\sqrt{-10b}} \sec h^2(kx), \tag{11}$$

where $k = \frac{1}{2}\sqrt{\frac{-a}{5b}}$.

The exact solution of mKDE is [1, 13]

$$u(x,\, t) = \frac{3a}{\sqrt{-10b}} \sec h^2(k(x - ct)), \tag{12}$$

where $c = \frac{25b-4a^2}{\sqrt{25b}}$.

Taking Laplace transform on Eq. (1),

$$s\mathcal{L}[u(x,t)] - u(x,\, 0) = -\mathcal{L}[u^2u_x] - a\mathcal{L}[u_{xxx}] - b\mathcal{L}[u_{xxxxx}] \tag{13}$$

$$\mathcal{L}[u(x,t)] = \frac{1}{s}\frac{3a}{\sqrt{-10b}} \sec h^2(kx) - \frac{1}{s}\{\mathcal{L}[u^2u_x] + a\mathcal{L}[u_{xxx}] + b\mathcal{L}[u_{xxxxx}]\}. \tag{14}$$

Taking inverse Laplace transform we get

$$u(x,t) = \frac{3a}{\sqrt{-10b}} \sec h^2(kx) - \mathcal{L}^{-1}\left\{\frac{1}{s}\{\mathcal{L}[u^2u_x] + a\mathcal{L}[u_{xxx}] + b\mathcal{L}[u_{xxxxx}]\}\right\}. \tag{15}$$

As described earlier, next we use HPM from (9):

$$\sum_{n=0}^{\infty} p^n u_n(x,t) = \frac{3a}{\sqrt{-10b}} \sec h^2(kx) - p\left(\mathcal{L}^{-1}\left[\frac{1}{s}\mathcal{L}\left[R\sum_{n=0}^{\infty} p^n u_n(x,t)\right] + \frac{1}{s}\mathcal{L}\left[\sum_{n=0}^{\infty} p^n H_n(u)\right]\right]\right). \tag{16}$$

Here, $R\sum_{n=0}^{\infty} p^n u_n(x,t)$ is related to linear terms u_{xxx} and u_{xxxxx}, whereas $\sum_{n=0}^{\infty} p^n H_n(u)$ represents the nonlinear term u^2u_x.

Comparing the coefficients of like powers of "p" on both sides of (16), we may obtain the following successive approximations:

$$p^0 : u_0(x,t) = \frac{3a}{\sqrt{-10b}} \sec h^2(kx)$$

$$p^1 : u_1(x,t) = -\mathcal{L}^{-1}\left[\frac{1}{s^2}\mathcal{L}[Ru_0(x,t)] + \frac{1}{s^2}\mathcal{L}[H_0(u)]\right],$$

$$Ru_0(x,t) = a(u_0(x,t))_{xxx} + b(u_0(x,t))_{xxxxx},$$

$$H_0(u) = [u_0(x,t)]^2(u_0(x,t))_x$$

$$p^2 : u_2(x,t) = -\mathcal{L}^{-1}\left[\frac{1}{s^2}\mathcal{L}[Ru_1(x,t)] + \frac{1}{s^2}\mathcal{L}[H_1(u)]\right],$$

$$Ru_1(x,t) = a(u_1(x,t))_{xxx} + b(u_1(x,t))_{xxxxx},$$

$$H_1(u) = 2u_0(x,t)u_1(x,t)(u_0(x,t))_x + [u_0(x,t)]^2(u_1(x,t))_x$$

$$p^3 : u_3(x,t) = -\mathcal{L}^{-1}\left[\frac{1}{s^2}\mathcal{L}[Ru_2(x,t)] + \frac{1}{s^2}\mathcal{L}[H_2(u)]\right],$$

$$Ru_2(x,t) = a(u_2(x,t))_{xxx} + b(u_2(x,t))_{xxxxx},$$

$$H_2(u) = (2\{u_1(x,t)\}^2 u_0(x,t)u_2(x,t))(u_0(x,t))_x + 2u_0(x,t)u_1(x,t)(u_1(x,t))_x + [u_0(x,t)]^2(u_2(x,t))_x$$

$$\vdots$$

The solution of mKDE (1) may be obtained as

$$u(x,\, t) = u_0(x,t) + u_1(x,t) + u_2(x,t) + u_3(x,t) + \cdots . \tag{17}$$

4. Solution of Interval-Modified Kawahara Equation

In this section, the coefficient of the third term u_x in mKDE, "a" (Eq. (1)), is taken as the interval for the interval-modified Kawahara differential equation (ImKDE) as follows:

$$\tilde{u}_t + \tilde{u}^2\tilde{u}_x + \tilde{a}\tilde{u}_{xxx} + b\tilde{u}_{xxxxx} = 0, \tag{18}$$

where "$\sim$" denotes interval uncertainty:

$$\tilde{u} = [\underline{u}, \overline{u}],$$
$$\tilde{a} = [\underline{a}, \overline{a}],$$

$\underline{a}$ is the lower bound of $\tilde{a}$, $\overline{a}$ is the lower bound of $\tilde{a}$,
$\underline{u}$ is the lower bound of $\tilde{u}$, $\overline{u}$ is the lower bound of $\tilde{u}$.

Again, HPTM has been used to handle ImKDE (Eq. (18)) to find the lower- and upper-bound solutions of the interval solution $\tilde{u}$.

ImKDE (18) may be transformed into a parametric form with the help of a single parametric form. An interval $\tilde{a} = [\underline{a}, \overline{a}]$ can be converted to crisp form using single-parametric concept as [14, 15]

$$\tilde{a} = \underline{a} + 2\beta\Delta\tilde{a}, \tag{19}$$

where $\beta \in [0, 1]$ is a parameter and $\Delta\tilde{a} = \frac{(\overline{a} - \underline{a})}{2}$.

Using (19) in ImKDE (18), the parametric differential equation may be obtained as

$$\tilde{u}_t + \tilde{u}^2\tilde{u}_x + (\underline{a} + 2\beta\Delta\tilde{a})\tilde{u}_{xxx} + b\tilde{u}_{xxxxx} = 0. \tag{20}$$

For simplicity, assume $\underline{a} + 2\beta\Delta\tilde{a} = \lambda$,

$$\tilde{u}_t + \tilde{u}^2\tilde{u}_x + \lambda\tilde{u}_{xxx} + b\tilde{u}_{xxxxx} = 0. \tag{21}$$

Equation (21) may be handled easily as given in the crisp case.

5. Numerical Results

In this section, numerical results of the crisp and interval–modified Kawahara equation are presented. Numerical solution of (1) for $a = 0.0001$ and $b = -1$ is compared with the exact solution (12) in Figure 1 at $t = 5$. The obtained lower and upper bounds of the interval solution of ImKDE are presented in Figure 2. Here, the interval is taken as $\tilde{a} = [-0.00005, 0.00005]$.

One can see from Figure 1 that the exact and numerical solutions using HPTM are same. From Figure 2, it is clear that the center solution is the same as the crisp exact solution.

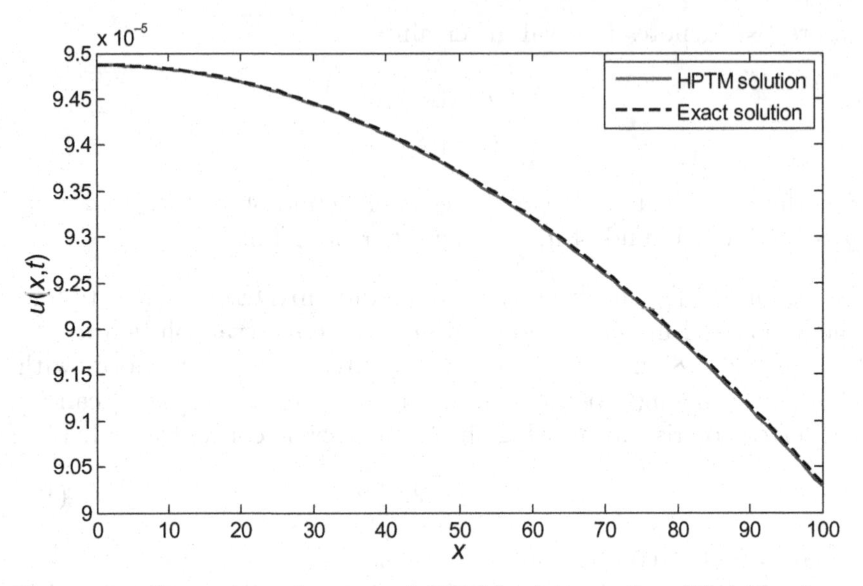

Figure 1. Comparison of exact and HPTM solutions of modified Kawahara equation at $t = 0.5$.

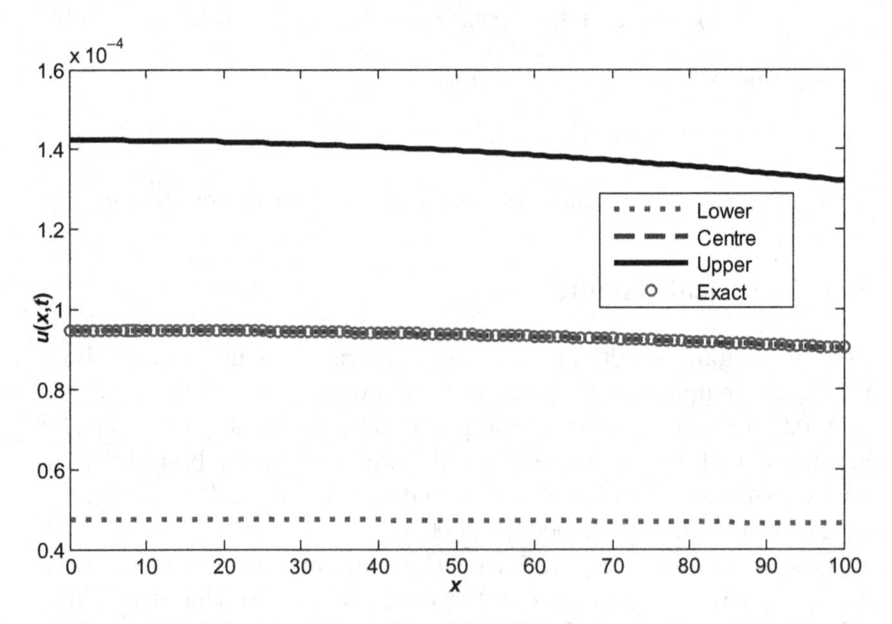

Figure 2. Solution of interval-modified Kawahara equation at $t = 0.5$.

6. Conclusion

HPTM has been successfully used to the interval–modified Kawahara differential equation to obtain the lower and upper bounds of the interval solution. The graphical results confirm that the exact and numerical solutions are in good agreement.

References

[1] Sirendaoreji (2004). New exact travelling wave solutions for the Kawahara and modified Kawahara equations. *Chaos Solitons & Fractals*, 19(1), 147–150.

[2] Jabbari, A. and Kheiri, H. (2010). New exact traveling wave solutions for the Kawahara and modified Kawahara equations by using modified tanh–coth method. *Acta Universitatis Apulensis*, 23, 21–38.

[3] Wazwaz, A. M. (2007). New solitary wave solutions to the modified Kawahara equation. *Physics Letters A*, 360(4–5), 588–592.

[4] Kurulay, M. (2012). Approximate analytic solutions of the modified Kawahara equation with homotopy analysis method. *Advances in Difference Equations*, 2012(1), 1–6.

[5] Abbasbandy, S. (2010). Homotopy analysis method for the Kawahara equation. *Nonlinear Analysis: Real World Applications*, 11(1), 307–312.

[6] Kashkari, B. S. (2014). Application of optimal homotopy asymptotic method for the approximate solution of Kawahara equation. *Applied Mathematical Sciences*, 8(18), 875–884.

[7] Wang, Q. (2011). The optimal homotopy-analysis method for Kawahara equation. *Nonlinear Analysis: Real World Applications*, 12(3), 1555–1561.

[8] Kawahara, T. (1972). Oscillatory solitary waves in dispersive media. *Journal of the Physical Society of Japan*, 33(1), 260–264.

[9] Karunakar, P. and Chakraverty, S. (2019). Solutions of time-fractional third- and fifth-order Korteweg–de-Vries equations using homotopy perturbation transform method. *Engineering Computations*, 36(7), 2309–2326.

[10] Chakraverty, S., Mahato, N., Karunakar, P., and Rao, T. D. (2019). *Advanced Numerical and Semi-Analytical Methods for Differential Equations*, Wiley, USA.

[11] Khan, M., Gondal, M. A., Hussain, I., and Vanani, S. K. (2012). A new comparative study between homotopy analysis transform method and homotopy perturbation transform method on a semi infinite domain. *Mathematical and Computer Modelling*, 55(3–4), 1143–1150.

[12] Gupta, S., Kumar, D., and Singh, J. (2015). Analytical solutions of convection–diffusion problems by combining Laplace transform method and homotopy perturbation method. *Alexandria Engineering Journal*, 54(3), 645–651.

[13] Jin, L. (2009). Application of variational iteration method and homotopy perturbation method to the modified Kawahara equation. *Mathematical and Computer modelling*, 49(3–4), 573–578.

[14] Karunakar, P. and Chakraverty, S. (2019). Effect of Coriolis constant on geophysical Korteweg-de Vries equation. *Journal of Ocean Engineering and Science*, 4(2), 113–121.

[15] Karunakar, P. and Chakraverty, S. (2018). Solution of interval shallow water wave equations using homotopy perturbation method. *Engineering Computations*, 35(4), 1610–1624.

https://doi.org/10.1142/9789811245367_0010

Chapter 10

Natural Convection of Sodium Alginate and Copper Nanofluid Inside Parallel Plates with Uncertain Parametric Behavior

P. S. Sangeetha* and Sukanta Nayak†,‡

Department of Mathematics,
Amrita School of Engineering,
Amrita Vishwa Vidyapeetham, Coimbatore, India

†*Department of Mathematics,*
School of Advanced Sciences, VIT-AP University,
Amaravati, Andhra Pradesh, India

‡*sukantgacr@gmail.com*

Abstract

This chapter explores the natural convection of the non-Newtonian fluid, sodium alginate, contaminated with copper nanoparticles between parallel plates. The same is considered with imprecisely defined parameters and investigated. The imprecisely defined parameters are taken as intervals and the said problem is studied using interval homotopy perturbation method. A novel percentage-parametric technique is introduced to handle the imprecise parameters with homotopy perturbation method and the desired uncertain solution is obtained.

Keywords: uncertainty, interval, homotopy perturbation method (HPM), interval homotopy perturbation method (IHPM), percentage-parametric technique, convection

1. Introduction

Natural convection heat transfer is a common phenomenon which comes into play in various physical and engineering applications, *viz.* heat exchangers, petroleum reservoirs, nuclear waste management, geothermal systems, and chemical catalytic reactors [1–4]. In view of this, many researchers have reported a number of articles based on the flow of fluids, and in particular, Newtonian and non-Newtonian fluid flows through two infinite parallel vertical plates. Bruce and Na [5] investigated the natural convection of non-Newtonian fluids between vertical flat plates. Ziabakhsh and Domairry [6] studied laminar natural convection problems. Whereas, Pawar *et al.* [7] performed experiments on isothermal steady state and non-isothermal unsteady state conditions in helical coils for Newtonian (water, glycerol–water mixture) and non-Newtonian (dilute aqueous polymer solutions of sodium carboxy methyl cellulose (SCMC), sodium alginate (SA) mixture) fluids. Tang *et al.* [8] investigated the nature of the flow of deionized water and the PAM solution for a range of Reynolds numbers as well as the non-Newtonian fluid flow in a microchannel. Yashino *et al.* [9] used lattice Boltzmann method to study incompressible non-Newtonian fluid flows. Further, in this broad area, Xu and Liao [10] studied the unsteady magnetohydrodynamic viscous flows of non-Newtonian fluids caused by an impulsively stretching plate by means of homotopy analysis method to investigate the effect of integral power-law index of these non-Newtonian fluids on the velocity. In another experimental work, Hojjat *et al.* [11] prepared three different types of nanofluids by dispersing Al_2O_3, TiO_2, and CuO nanoparticles in a 0.5 wt% of carboxymethyl cellulose (CMC) aqueous solution. They measured the thermal conductivity of the base fluid and nanofluids with various nanoparticle loadings at different temperatures. Their results show that the thermal conductivity of the nanofluids is higher than that of the base fluid, and the increase in the thermal conductivity varies exponentially with the nanoparticle concentration [11]. There are some simple and accurate approximation techniques for solving differential equations called the weighted residuals methods (WRMs). Collocation, Galerkin, and least square are examples of WRMs. Stern and Rasmussen [12] used collocation method for solving a third-order linear differential equation. Vaferi *et al.* [13] studied the feasibility of applying orthogonal collocation method to solve the diffusivity equation in a radial transient flow system. Hendi and Albugami [14] used collocation and Galerkin methods for solving the Fredholm–Volterra integral equation. Further, Aziz and Bouaziz [15] used least-square method for predicting the performance of a longitudinal fin. They found that the least-squares method is simple compared with other analytical methods. Shaoqin and Huoyuan [16] developed and analyzed least-squares approximations for the incompressible magnetohydrodynamic equations. The concept of differential transformation method (DTM) was first introduced by Zhou [17] in 1986, and it was used to solve both linear and nonlinear initial-value problems in electric circuit analysis. This method can be applied directly for linear and nonlinear differential equations without requiring linearization, discretization, or perturbation, and this is the main benefit of this method. Ghafoori *et al.* [18] used the DTM for solving linear oscillation equation. Other analytical studies in this field can be found in [19–21].

In the literature reviewed, the wide range of application as well as the importance is noticeable. As such, various researchers employed many methods, and it is found that the homotopy perturbation method (HPM) is simple to use. Besides this, the presence of uncertainties makes the problem complicated but more realistic. Hence, the said problem is investigated with the presence of uncertainties. The uncertainties occur due to impreciseness, vagueness, and lack of information about the involved parameters, constants, coefficients, and experimental values. In this regard, many researchers used various techniques to handle these uncertainties effectively. Some of the research works are discussed here to understand the challenges involved and the need to handle them.

Qiu *et al.* [22] used an interval approach in reliability theory and obtained an interval probability of system failure. The combination of reliability theory and interval analysis is then applied to solve two example problems of truss systems. Whereas, Blackwell and Beck [23] considered thermal properties and temperature measurements as uncertainties to analyze linear inverse heat-conduction problem to estimate heat flux from interior temperatures. Chakraverty and Nayak [24] studied in detail the interval uncertainties and redefined interval arithmetic through a transformation that converts intervals into crisp form by symbolic parametrization. Then, the interval arithmetic is extended to fuzzy numbers, and the fuzzy finite-element method is developed. The same is used to solve a heat-conduction problem. Besides, Nayak and Chakraverty [25] employed modified fuzzy finite-element method (FFEM), where they changed α-cut interval fuzzy numbers into crisp form, to investigate the sensitiveness of the uncertain parameters involved in conjugate heat transfer in a square plate. Further, Nayak *et al.* [26] studied the distribution of temperatures in a conduction–convection system under uncertain environments. With modification in the representation of intervals, Wang and Qiu [27] introduced an interval parameter-perturbation method (IPPM) and a sensitivity-based interval analysis method (SIAM). Then, the same are used to estimate temperatures for a steady-state heat convection–diffusion problem with uncertainties in material properties, external loads, and boundary conditions. Li *et al.* [28] presented several numerical algorithms to solve the convection–diffusion equation with random diffusivity and periodic boundary conditions. In this respect, it is observed that there is a need for an alternative tool to handle interval uncertainties with the occurrence of percentage errors. Hence, a novel technique of percentage-parametric approach is proposed to compute intervals. Further, this technique is used to study the natural convection of a non-Newtonian fluid, sodium alginate, contaminated with copper nanoparticles between parallel plates considering uncertainties.

2. Natural Convection of Nanofluid Inside Parallel Plates

A schematic of a non-Newtonian fluid flow between two vertical plates is shown in Figure 1. The problem consists of two vertical plates, which are separated by a distance $2b$. In between the two vertical parallel plates, a non-Newtonian fluid flows with natural convection. Constant temperatures T_2 and T_1 are maintained at the plate walls, i.e., at $x = b$ and $x = -b$, respectively, where $T_1 > T_2$. This difference in temperature causes the fluid near the wall at $x = b$ to fall and the fluid near the wall at $x = -b$ to rise. The fluid is a non-Newtonian sodium alginate–based nanofluid containing Cu and Ag nanoparticles. It is assumed that the base fluid and the nanoparticles are in thermal equilibrium and no slip occurs between

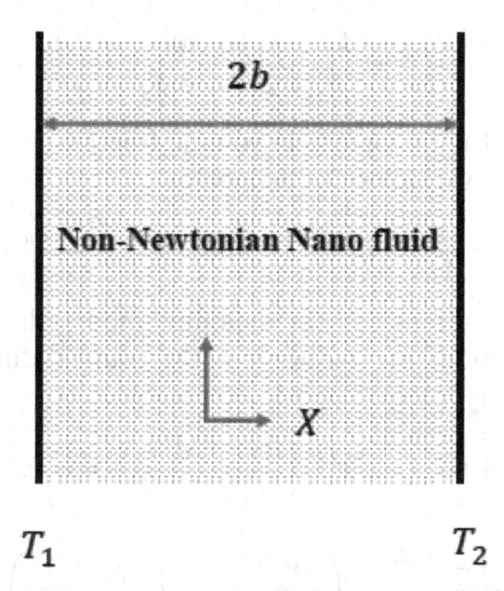

Figure 1. Diagrammatic representation of a non-Newtonian nanofluid flow between two vertical flat plates.

Table 1. Thermophysical properties of non-Newtonian fluid and nanoparticles.

Material	Density (kg/m^3)	Specific Heat, C_p(J/kgK)	Thermal Conductivity (W/mK)
Copper (Cu)	8933	385	401
Silver (Ag)	10500	235	429
Sodium Alginate (SA)	989	4175	0.6376

them. The thermophysical properties of the nanofluid are given in Table 1.

The effective density, effective dynamic viscosity, heat capacitance, and thermal conductivity of the nanofluid are given as [1, 2]

$$\rho_{nf} = \rho_f(1-\emptyset) + \rho_s\emptyset, \tag{1}$$

$$\mu_{nf} = \frac{\mu_f}{(1-\emptyset)^{2.5}}, \tag{2}$$

$$(\rho C_p)_{nf} = (\rho C_p)_f(1-\emptyset) + (\rho C_p)_s\emptyset, \tag{3}$$

$$\frac{k_{nf}}{k_f} = \frac{k_s + 2k_f + 2\emptyset(k_s - k_f)}{k_s + 2k_f - \emptyset(k_s - k_f)}, \quad (4)$$

where $\emptyset$ denotes the nanoparticle volume fraction.

The similarity variables are defined as

$$V = \frac{v}{V_0}, \quad \eta = \frac{x}{b}, \quad \text{and} \quad \theta = \frac{T - T_m}{T_1 - T_2}. \quad (5)$$

Using these assumptions and the nanofluid model, the Navier–Stokes and energy equations are reduced to the coupled-differential equations as

$$\frac{d^2V}{d\eta^2} + 6\delta\,(1-\emptyset)^{2.5}\left(\frac{dV}{d\eta}\right)^2 \frac{d^2V}{d\eta^2} + \theta = 0, \quad (6)$$

$$\frac{d^2\theta}{d\eta^2} + EcPr\left(\frac{(1-\emptyset)^{2.5}}{A_1}\right)\left(\frac{dV}{d\eta^2}\right)^2 + 2\delta EcPr\left(\frac{1}{A_1}\right)\left(\frac{dV}{d\eta}\right)^4, \quad (7)$$

where Pr and Ec stand for Prandtl and Eckert numbers, respectively, δ for dimensionless non-Newtonian viscosity, A_1 for ratio of the thermal conductivity of nanofluid and that of the real fluid, and they have the following forms:

$$Ec = \frac{\rho_f {V_0}^2}{(\rho C_p)_f(\theta_1 - \theta_2)}, \quad Pr = \frac{\mu_f(\rho C_p)_f}{\rho_f K_f}, \quad \delta = \frac{6\beta_3 V_0^2}{\mu_f b^2}, \quad (8)$$

$$A_1 = \frac{k_{nf}}{k_f} = \frac{k_s + 2k_f + 2\emptyset(k_s - k_f)}{k_s + 2k_f - \emptyset(k_s - k_f)}. \quad (9)$$

The boundary conditions are taken as

$$V(\eta) = \begin{cases} 0, & \eta = 1 \\ 0, & \eta = -1 \end{cases}, \quad (10)$$

$$\theta(\eta) = \begin{cases} -0.5, & \eta = 1 \\ 0.5, & \eta = -1 \end{cases}. \quad (11)$$

It is worth mentioning that even a small change in the value of nanoparticle volume fraction can affect the velocity and temperature profiles. Now, we handle such a problem in an uncertain environment using interval analysis.

3. Interval Homotopy Perturbation Method (IHPM)

Here, the concept of interval analysis is hybridized with the homotopy perturbation method (HPM) to develop IHPM. The uncertainties involved are computed through the interval computation technique. To understand the same, first the interval arithmetic is discussed and then, the HPM is included.

An interval A can be defined as [29]

$$A = [a_1, a_2] = \{t | a_1 \leq t \leq a_2, a_1, a_2 \in \mathcal{R}\}, \tag{12}$$

where a_1 and a_2 are the left and right endpoints of the interval A, respectively.

Two intervals are said to be equal if the left endpoints and right endpoints for both the intervals are the same. For example, if we take two intervals, *viz.* $A = [a_1, a_2]$ and $B = [b_1, b_2]$, then they are said to be equal if $a_1 = b_1$ and $a_2 = b_2$.

The width of an interval A is defined by $w = a_2 - a_1$. Whereas, the midpoint of A is $m = \frac{a_1+a_2}{2}$.

To compute interval uncertainties, the traditional interval arithmetic operations are defined as

$$[a_1, a_2] + [b_1, b_2] = [a_1 + b_1, a_2 + b_2], \tag{13}$$

$$[a_1, a_2] - [b_1, b_2] = [a_1 - b_2, a_2 - b_1], \tag{14}$$

$$[a_1, a_2] \cdot [b_1, b_2] = [\min(a_1 b_1, a_1 b_2, a_2 b_1, a_2 b_2), \times \max(a_1 b_1, a_1 b_2, a_2 b_1, a_2 b_2)], \tag{15}$$

$$\frac{[a_1, a_2]}{[b_1, b_2]} = \left[\min\left(\frac{a_1}{b_1}, \frac{a_1}{b_2}, \frac{a_2}{b_1}, \frac{a_2}{b_2}\right), \max\left(\frac{a_1}{b_1}, \frac{a_1}{b_2}, \frac{a_2}{b_1}, \frac{a_2}{b_2}\right)\right],$$

where $b_1, b_2 \neq 0$. (16)

Alternatively, an interval $A = [a_1, a_2]$ can be written as the following crisp form [30, 31] by using a parameter $\alpha, 0 \leq \alpha \leq 1$:

$$A = f(\alpha) = a_1 + \alpha(a_2 - a_1). \tag{17}$$

Substituting $\alpha = 0$ and $\alpha = 1$, one may get the lower and upper bounds of the interval A, respectively. The expression in Eq. (17) converts the interval into a single-variable function, which is used here to compute the interval uncertainties.

Further, the need for handling percentage error and interval uncertainties motivates us to propose an alternative technique to compute intervals effectively. In this regard, in the following, a percentage parametric technique is discussed.

For the interval $[a, b]$ the midpoint is defined as $m = \frac{a+b}{2}$, and Δ is the percentage of error. The width of the interval is $w = b - a$. Then, there exist a number

$$x = a + \frac{\beta}{\Delta} w, \tag{18}$$

where $\beta \epsilon [0, \Delta]$.

If Δ is considered to be 5% of $m = 10$, then the interval is $[9.5, 10.5]$, where $\beta \epsilon [0, 0.5]$. The interval $x = [a, b]$ can be transformed into the following crisp form:

$$f(\beta) = a + \frac{\beta}{\Delta} w, \tag{19}$$

where β is a parameter lying in the closed interval $[0, 0.5]$.

So, Eq. (19) can be written as $f(\beta) = \frac{-1}{\Delta}(\Delta^2 - m\Delta - \beta w) = \left(9.5 + \frac{\beta}{0.5}(b - a)\right)$.

The basic idea of HPM can be understood from the following differential equation.

Let us consider the differential equation

$$A(u) - f(r) = 0, \quad r \in \Omega, \tag{20}$$

with the boundary condition

$$B\left(u, \frac{dV}{d\eta}\right) = 0, \quad r \in \Gamma, \tag{21}$$

where A is the differential operator, B is the boundary operator, $f(r)$ is a known analytical function, and γ is the boundary of the domain Ω.

The operator A can be classified into two parts, *viz.* linear (L) and nonlinear (N). Then, Eq. (20) can be written as

$$L(u) + N(u) - f(r) = 0, \quad r \in \Omega. \tag{22}$$

One may construct a homotopy $v(r, q) : \Omega \times [0, 1] \rightarrow R$ satisfying

$$H(v, q) = (1 - q)[L(v) - L(u_0)] + q[A(v) - f(r)] = 0, \tag{23}$$

where q is an embedding parameter lying between 0 and 1, u_0 is an initial approximation satisfying the boundary condition in Eq. (21).

From Eq. (23), we observe that when $q = 0, L(v) = L(u_0)$ and for $q = 1, A(v) - f(r) = 0$.

In other words, when q converges to 1, we get the solution of Eq. (20). As q is a small parameter, the solution of Eq. (23) can be expressed in terms of a power series in q:

$$v = v_0 + qv_1 + q^2 v_2 + q^3 v_3 + \cdots. \tag{24}$$

By setting $q = 1$, we get the solution of Eq. (20) as

$$v = \lim_{q \to 1} (v_0 + qv_1 + q^2 v_2 + q^3 v_3 + \cdots). \tag{25}$$

In this chapter, we have hybridized the concept of interval uncertainties with the usual HPM to get IHPM. In the following sections, the nanofluid heat-transfer problem is stated and the same is solved by using IHPM.

4. Solution of Natural Convection of Sodium Alginate and Copper Nanofluid Inside Parallel Plates using Percentage-Parametric HPM

Considering the percentage-parametric technique, the coupled differential equations (Eqs. (6) and (7)) in interval form can be written as

$$\begin{aligned}&\left(9.5 + \frac{\beta}{0.5}\left(\overline{v}'' - \underline{v''}\right)\right) + 6\delta\left(1 - \left(9.5 + \frac{\beta}{0.5}(\overline{\emptyset} - \underline{\emptyset})\right)\right)^{2.5}\\&\quad \times \left(9.5 + \frac{\beta}{0.5}\left(\overline{v}' - \underline{v'}\right)\right)^2 \left(9.5 + \frac{\beta}{0.5}\left(\overline{v}'' - \underline{v''}\right)\right)\\&\quad + \left(9.5 + \frac{\beta}{0.5}\left(\overline{\theta} - \underline{\theta}\right)\right) = 0,\end{aligned} \tag{26}$$

$$\left(9.5+\frac{\beta}{0.5}(\overline{\theta}''-\underline{\theta''})\right)+Ec.\Pr\left(\frac{\left(1-\left(9.5+\frac{\beta}{0.5}(\overline{\emptyset}-\underline{\emptyset})\right)\right)^{2.5}}{\left(9.5+\frac{\beta}{0.5}\left(\overline{A}_1-\underline{A_1}\right)\right)}\right)$$

$$\times\left(9.5+\frac{\beta}{0.5}\left(\overline{v}'-\underline{v}'\right)\right)^2+2\delta Ec.Pr\frac{1}{\left(9.5+\frac{\beta}{0.5}\left(\overline{A}_1-\underline{A_1}\right)\right)}$$

$$\times\left(9.5+\frac{\beta}{0.5}\left(\overline{v}'-\underline{v}'\right)\right)^4=0. \tag{27}$$

For simplicity, we may write

$$9.5+\frac{\beta}{0.5}(\overline{v}''-\underline{v''})=f_v''(\eta,\beta)$$

$$9.5+\frac{\beta}{0.5}(\overline{\emptyset}-\underline{\emptyset})=\emptyset(\beta)$$

$$9.5+\frac{\beta}{0.5}(\overline{\theta}-\underline{\theta})=f_\theta(\eta,\beta).$$

Then, Eqs. (26) and (27) become

$$f_v''(\eta,\beta)+6\delta(1-\emptyset(\beta))^{2.5}(f_v'(\eta,\beta))^2f_v''(\eta,\beta)+f_\theta(\eta,\beta)=0, \tag{28}$$

$$f_\theta''(\eta,\beta)+Ec.\Pr\left(\frac{(1-\emptyset(\beta))^{-2.5}}{A_1(\beta)}\right)f_v'(\eta,\beta)$$

$$+2\delta Ec.\,Pr.\left(\frac{1}{A_1(\beta)}\right)(f_v'(\eta,\beta))^4=0. \tag{29}$$

Similarly, the boundary conditions can be expressed as

$$f_v(\eta,\beta)=\begin{cases}0, & \eta=1\\ 0, & \eta=-1\end{cases}, \tag{30}$$

$$f_\theta\,(\eta,\beta)=\begin{cases}-0.5, & \eta=1\\ 0.5, & \eta=-1\end{cases}. \tag{31}$$

Now, we apply HPM to solve Eqs. (28) and (29). Homotopy for Eqs. (28) and (29) is constructed as

$$H_1(q,\eta)=(1-q)[f_v''(\eta,\beta)+f_\theta(\eta,\beta)]+q[f_v''(\eta,\beta)$$

$$+6\delta(1-\emptyset(\beta))^{2.5}(f_v'(\eta,\beta))^2f_v''(\eta,\beta)+f_\theta(\eta,\beta)]=0, \tag{32}$$

$$H_2(q,\eta) = (1-q)[f''_\theta(\eta,\beta)] + q\left[f''_\theta(\eta,\beta) + Ec.Pr.\left(\frac{(1-\emptyset(\beta))^{-2.5}}{A_1(\beta)}\right)\right.$$
$$\left.\times(f'_v(\eta,\beta))^2 + 2\delta Ec.Pr.\left(\frac{1}{A_1(\beta)}\right)(f'_v(\eta,\beta))^4\right] = 0. \tag{33}$$

According to the percentage-parametric technique and HPM,

$$f_v(\eta,\beta) = f_{v0}(\eta,\beta) + qf_{v_1}(\eta,\beta) + q^2 f_{v_2}(\eta,\beta) + q^3 f_{v_3}(\eta,\beta) + \cdots \tag{34}$$

and

$$f_\theta(\eta,\beta) = f_{\theta_0}(\eta,\beta) + qf_{\theta_1}(\eta,\beta) + q^2 f_{\theta_2}(\eta,\beta) + q^3 f_{\theta_3}(\eta,\beta) + \cdots \tag{35}$$

are the assumed series solution of Eqs. (24) and (25).

The aim is to find the unknown functions $f_{v0}(\eta,\beta)$, $f_{v1}(\eta,\beta)$, $f_{v_2}(\eta,\beta)$, $f_{v_3}(\eta,\beta)$, and so on for velocity profile and $f_{\theta_0}(\eta,\beta)$, $f_{\theta_1}(\eta,\beta)$, $f_{\theta_2}(\eta,\beta)$, $f_{\theta_3}(\eta,\beta)$, and so on for temperature profile.

On assuming that solutions $f_v(\eta,\beta)$ and $f_\theta(\eta,\beta)$ should satisfy the differential equations by substituting Eqs. (26) and (27) in Eqs. (24) and (25) and collecting the various powers of q, we get

$$[f''_{v0}(\eta,\beta) + f_{\theta_0}(\eta,\beta)]q^0 + [f''_{v_1}(\eta,\beta) + 6\delta(1-\emptyset(\beta))^{2.5}$$
$$\times(f'_{v0}(\eta,\beta))^2 f''_{v0}(\eta,\beta) + f_{\theta_1}(\eta,\beta)]q^1 + [f''_{v2}(\eta,\beta)$$
$$+6\delta(1-\emptyset(\beta))^{2.5}(f'_{v_1}(\eta,\beta))^2 f''_{v_1}(\eta,\beta) + f_{\theta_2}(\eta,\beta)]q^2 + \cdots = 0, \tag{36}$$

$$[f''_{\theta_0}(\eta,\beta)]q^0 + [f''_{\theta_1}(\eta,\beta) + Ec.Pr.\left(\frac{(1-\emptyset(\beta))^{-2.5}}{A_1(\beta))}\right)(f'_{v0}(\eta,\beta))^2$$
$$+2\delta Ec.Pr.\left(\frac{1}{A_1(\beta)}\right)(f'_{v0}(\eta,\beta))^4]q^1$$
$$+\left[f''_{\theta_2}(\eta,\beta) + Ec.Pr.\left(\frac{(1-\emptyset(\beta))^{-2.5}}{A_1(\beta)}\right)(f'_{v_1}(\eta,\beta))^2\right.$$
$$\left.+2\delta Ec.Pr.\left(\frac{1}{A_1(\beta)}\right)(f'_{v_1}(\eta,\beta))^4\right]q^2 + \cdots = 0. \tag{37}$$

Equating the coefficients of various powers of q from Eqs. (36) and (37) to zero separately and also by using boundary conditions, we get the functions $f_{v_0}(\eta,\beta)$, $f_{v_1}(\eta,\beta)$, $f_{v_2}(\eta,\beta)$, $f_{v_3}(\eta,\beta)$, and so on for velocity profile and $f_{\theta_0}(\eta,\beta)$, $f_{\theta_1}(\eta,\beta)$, $f_{\theta_2}(\eta,\beta)$, $f_{\theta_3}(\eta,\beta)$, and so on for temperature profile.

First, equating the coefficient of q^0 in Eq. (37) to zero, we get

$$f''_{\theta_0}(\eta,\beta) = 0, \tag{38}$$

with the boundary conditions

$$f_{\theta_0}(\eta,\beta) = \begin{cases} -0.5, & \eta = 1 \\ 0.5, & \eta = -1 \end{cases}. \tag{39}$$

From Eqs. (38) and (39), we have

$$f_{\theta_0}(\eta,\beta) = \frac{-\eta}{2}. \tag{40}$$

Next, equating the coefficient of q^0 in Eq. (36) to zero, we obtain

$$f''_{v_0}(\eta,\beta) + f_{\theta_0}(\eta,\beta) = 0, \tag{41}$$

with boundary conditions

$$f_{v_0}(\eta,\beta) = \begin{cases} 0, & \eta = 1 \\ 0, & \eta = -1 \end{cases}. \tag{42}$$

From Eqs. (41) and (42), we get

$$f_{v_0}(\eta,\beta) = \frac{\eta^3}{12} - \frac{\eta}{12}. \tag{43}$$

First, equating the coefficient of q^1 in Eq. (37) to zero, we have

$$\begin{aligned} f''_{\theta_1}(\eta,\beta) &+ Ec.\,Pr.\left(\frac{(1-\emptyset(\beta))^{-2.5}}{A_1(\beta)}\right)(f'_{v_0}(\eta,\beta))^2 \\ &+ 2\delta Ec.\,Pr.\left(\frac{1}{A_1(\beta)}\right)(f'_{v_0}(\eta,\beta)t)^4 = 0, \end{aligned} \tag{44}$$

with boundary conditions

$$f_{\theta_1}(\eta,\beta) = \begin{cases} 0, & \eta = 1 \\ 0, & \eta = -1 \end{cases}. \tag{45}$$

From Eqs. (44) and (45), we get

$$f_{\theta_1} = \frac{-Ec.Pr.}{12}\left(\frac{(1-\emptyset(\beta))^{-2.5}}{A_1(\beta)}\right)\left[\frac{\eta^6}{40} - \frac{\eta^4}{24} + \frac{\eta^2}{24} - \frac{1}{40}\right]$$
$$-\frac{2\delta Ec.Pr.}{12^2}\left(\frac{1}{A_1(\beta)}\right)\left[\frac{\eta^{10}}{160} - \frac{3\eta^8}{224} + \frac{\eta^6}{80} - \frac{\eta^4}{144} + \frac{\eta^2}{288} - \frac{19}{10080}\right]. \quad (46)$$

First, equating the coefficient of q^1 in Eq. (46) to zero, we have

$$f''_{v_1}(\eta,\beta) + 6\delta(1-\emptyset(\beta))^{2.5}(f'_{v_0}(\eta,\beta))^2 f''_{v_0}(\eta,\beta) + f_{\theta_1}(\eta,\beta) = 0, \quad (47)$$

with boundary conditions

$$f_{v_1}(\eta,\beta) = \begin{cases} 0, & \eta = 1 \\ 0, & \eta = -1 \end{cases}. \quad (48)$$

From Eqs. (46) and (47), we get

$$f_{v_1} = \frac{-6\delta(1-\emptyset(\beta))^{2.5}}{12}\left[\frac{\eta^7}{112} - \frac{\eta^5}{80} + \frac{\eta^3}{144} - \frac{17\eta}{5040}\right]$$
$$-\frac{Ec.Pr.}{12}\left(\frac{(1-\emptyset(\beta))^{-2.5}}{A_1(\beta)}\right)$$
$$\times\left[\frac{\eta^8}{2240} - \frac{\eta^6}{1612800} + \frac{\eta^4}{288} - \frac{\eta^2}{80} + \frac{67}{6720}\right]$$
$$-\frac{2\delta Ec.Pr.}{12^3}\left(\frac{1}{A_1(\beta)}\right)\left[\frac{\eta^{12}}{1760} - \frac{3\eta^{10}}{1680} + \frac{3\eta^8}{1120} - \frac{\eta^6}{360} + \frac{\eta^4}{288}\right.$$
$$\left. - \frac{19\eta^2}{1680} + \frac{203}{22179}\right]. \quad (49)$$

By continuing the same process for different powers of q from Eqs. (36) and (37), we get $f_{\theta_2}(\eta,\beta)$, $f_{v_2}(\eta,\beta)$, $f_{\theta_3}(\eta,\beta)$, $f_{v_3}(\eta,\beta)$ and so on.

According to this method, the approximate solution for velocity and temperature profiles are given by

$$\begin{aligned} f_v(\eta,\beta) = {} & \frac{\eta^3}{12} - \frac{\eta}{12} - \frac{6\delta(1-\emptyset(\beta))^{2.5}}{12}\left[\frac{\eta^7}{112} - \frac{\eta^5}{80} + \frac{\eta^3}{144} - \frac{17\eta}{5040}\right] \\ & - \frac{Ec.Pr.}{12}\left(\frac{(1-\emptyset(\beta))^{-2.5}}{A_1(\beta)}\right) \\ & \times \left[\frac{\eta^8}{2240} - \frac{\eta^6}{1612800} + \frac{\eta^4}{288} - \frac{\eta^2}{80} + \frac{67}{6720}\right] \\ & - \frac{2\delta Ec.Pr.}{12^3}\left(\frac{1}{A_1(\beta)}\right)\left[\frac{\eta^{12}}{1760} - \frac{3\eta^{10}}{1680} + \frac{3\eta^8}{1120} - \frac{\eta^6}{360} + \frac{\eta^4}{288} \right. \\ & \left. - \frac{19\eta^2}{1680} + \frac{203}{22179}\right] \end{aligned} \tag{50}$$

and

$$\begin{aligned} f_\theta(\eta,\beta) = {} & \frac{-\eta}{2} - \frac{Ec.Pr.}{12}\left(\frac{(1-\emptyset(\beta))^{-2.5}}{A_1(\beta)}\right)\left[\frac{\eta^6}{40} - \frac{\eta^4}{24} + \frac{\eta^2}{24} - \frac{1}{40}\right] \\ & - \frac{2\delta Ec.Pr.}{12^2}\left(\frac{1}{A_1(\beta)}\right)\left[\frac{\eta^{10}}{160} - \frac{3\eta^8}{224} + \frac{\eta^6}{80} - \frac{\eta^4}{144} + \frac{\eta^2}{288} \right. \\ & \left. - \frac{19}{10080}\right], \end{aligned} \tag{51}$$

respectively.

5. Concluding Remarks

This chapter describes a novel percentage-parametric approach to handle uncertain values involved in a problem. The main advantage of this method is that the percentage of error in uncertainties are initially considered and computed by using parametric transformation. The percentage of error gives us freedom to choose the level of uncertainness that is tolerance of error in uncertain parameters. Also, this technique can convert the intervals to single-variable functions, which reduces the computation effort. Using the same, generalized series solution is obtained for the velocity and temperature profiles in natural convection of a non-Newtonian fluid (sodium alginate) contaminated with copper nanoparticles between parallel plates.

Nomenclature

C_p	Specific heat capacity
k_f	Thermal conductivity of the fluid
k_s	Thermal conductivity of the nanoparticles
k_{nf}	Thermal conductivity of the nanofluid
μ_f	Dynamic viscosity of the fluid
μ_{nf}	Effective dynamic viscosity of the nanofluid
ρ_f	Density of the fluid
ρ_s	Density of the nanoparticles
ρ_{nf}	Effective density of the nanofluid
$(\rho C_p)_f$	Thermal capacitance of the fluid
$(\rho C_p)_s$	Thermal capacitance of the nanoparticles
$(\rho C_p)_{nf}$	Heat capacitance of the nanofluid
$\emptyset$	Nanoparticle volume fraction
θ	Temperature
v	Volume

References

[1] Hatami, M. and Ganji, D. D. (2013). Heat transfer and flow analysis for SA-TiO2 non-Newtonian nanofluid passing through the porous media between two coaxial cylinders. *Journal of Molecular Liquids*, 188, 155–161.

[2] Hatami, M., Nouri, R., and Ganji, D. D. (2013). Forced convection analysis for MHD Al2O3-water nanofluid flow over a horizontal plate. *Journal of Molecular Liquids*, 187, 294–301.

[3] Sheikholeslami, M., Gorji-Bandpy, M., and Ganji, D. D. (2013). Natural convection in a nanofluid filled concentric annulus between an outer square cylinder and an inner elliptic cylinder. *Iranian Journal of Science and Technology, Transactions of Mechanical Engineering*, 20(4), 1241–1253.

[4] Sheikholeslami, M., Gorji-Bandpy, M., and Domairry, G. (2013). Free convection of nanofluid filled enclosure using Lattice Boltzmann Method (LBM). *Applied Mathematical Methods in Mechanical Engineering*, 34(7), 1–15.

[5] Bruce, R. W. and Na, T. Y. (1967). Natural convection flow of Powell–Eyring fluids between two vertical flat plates. Vol. 67, p. 25, ASME, New York.

[6] Ziabakhsh, Z. and Domairry, G. (2009). Analytic solution of natural convection flow of a non-Newtonian fluid between two vertical flat plates using homotopy analysis method. *Communications in Nonlinear Science and Numerical Simulation 1*, 14, 1868–1880.

[7] Pawar, S. S. and Sunnapwar Vivek, K. (2013). Experimental studies on heat transfer to Newtonian and non-Newtonian fluids in helical coils with laminar and turbulent flow. *Experimental Thermal and Fluid Science*, 44, 792–804.

[8] Tang, G. H., Lu, Y. B., Zhang, S. X., Wang, F. F, and Tao, W. Q. (2012). Experimental investigation of non-Newtonian liquid flow in microchannels. *Journal of Non-Newtonian Fluid Mechanics*, 173–174, 21–29.

[9] Yoshino, M., Hotta, Y., Hirozane, T., and Endo, M. (2007). A numerical method for incompressible non-Newtonian fluid flows based on the lattice Boltzmann method. *Journal of Non-Newtonian Fluid Mechanics*, 147, 69–78.

[10] Xu, H. and Liao, S. J. (2005). Series solutions of unsteady magneto hydrodynamic flows of non-Newtonian fluids caused by an impulsively stretching plate. *Journal of Non-Newtonian Fluid Mechanics*, 129, 46–55.

[11] Hojjat, M., Etemad, S. Gh., Bagheri, R., and Thibault, J. (2011). Thermal conductivity of non-Newtonian nanofluids: Experimental data and modeling using neural network. *International Journal of Heat and Mass Transfer*, 54, 1017–1023.

[12] Stern, R. H. and Rasmussen, H. (1996). Left ventricular ejection: Model solution by collocation, an approximate analytical method. *Computers in Biology and Medicine*, 26, 255–261.

[13] Vaferi, B., Salimi, V., Dehghan Baniani, D., Jahanmiri, A., and Khedri, S. (2012). Prediction of transient pressure response in the petroleum reservoirs using orthogonal collocation. *Journal of Petroleum Science and Engineering*, 98–99, 156–163.

[14] Hendi, F. A. and Albugami, A. M. (2010). Numerical solution for Fredholm–Volterra integral equation of the second kind by using collocation and Galerkin methods. *Journal of King Saud University — Science*, 22, 37–40.

[15] Aziz, A. and Bouaziz, M. N. (2011). A least squares method for a longitudinal fin with temperature dependent internal heat generation and thermal conductivity. *Energy Conversion and Management*, 52, 2876–2882.

[16] Shaoqin, G. and Huoyuan, D. (2008). Negative norm least-squares methods for the incompressible magneto-hydrodynamic equations. *Acta Mathematica Scientia*, 28B(3), 675–684.
[17] Zhou, J. K. (1986). *Differential Transformation Method and its Application for Electrical Circuits*, Hauzhang University Press: Wuhan, China.
[18] Ghafoori, S., Motevalli, M., Nejad, M. G., Shakeri, F., Ganji, D. D., and Jalaal, M. (2011). Efficiency of differential transformation method for nonlinear oscillation: Comparison with HPM and VIM. *Current Applied Physics*, 11, 965–971.
[19] Domairry, G. and Ziabakhsh, Z. (2009). Solution of the laminar viscous flow in a semi-porous channel in the presence of a uniform magnetic field by using the homotopy analysis method. *Communications in Nonlinear Science and Numerical Simulation*, 14(4), 1284.
[20] Ziabakhsh, Z., Domairry, G., and Esmaeilpour, M. (2009). Solution of the laminar viscous flow in a semi-porous channel in the presence of a uniform magnetic field by using the homotopy analysis method. *Communications in Nonlinear Science and Numerical Simulation*, 14, 1284–1294.
[21] Ziabakhsh, Z., Domairry, G., and Ghazizadeh, H. R. (2009). Analytical solution of the stagnation-point flow in a porous medium by using the homotopy analysis method. *Journal of the Taiwan Institute of Chemical Engineers*, 40, 91–97.
[22] Zhiping, Q. and Di Yang, I. E. (2008). Probabilistic interval reliability of structural systems. *International Journal of Solids and Structures*, 45, 2850–2860.
[23] Blackwell, B. and Beck, J. V. (2010). A technique for uncertainty analysis for inverse heat conduction problems. *International Journal of Heat and Mass Transfer*, 53, 753–759.
[24] Chakraverty, S. and Nayak, S. (2012). Fuzzy finite element method for solving uncertain heat conduction problems. *Coupled Systems Mechanics*, 1, 345–360.
[25] Nayak, S. and Chakraverty, S. (2013). Non-probabilistic approach to investigate uncertain conjugate heat transfer in an imprecisely defined plate. *International Journal of Heat and Mass Transfer*, 67, 445–454.
[26] Nayak, S., Chakraverty, S., and Datta, D. (2014). Uncertain spectrum of temperatures in a non homogeneous fin under imprecisely defined conduction-convection system. *Journal of Uncertain Systems*, 8, 123–135.
[27] Wang, C. and Qiu, Z. (2015). Interval analysis of steady-state heat convection–diffusion problem with uncertain-but-bounded parameters. *International Journal of Heat and Mass Transfer*, 91, 355–362.

[28] Li, N., Zhao, J., Feng, X., and Gui, D. (2016). Generalized polynomial chaos for the convection diffusion equation with uncertainty. *International Journal of Heat and Mass Transfer*, 97, 289–300.

[29] Nayak, S. and Chakraverty, S. (2018). Non-probabilistic solution of moving plate problem with uncertain parameters. *Journal of Fuzzy Set Valued Analysis*, 2, 49–59.

[30] Nayak, S. and Chakraverty, S. (2018). *Interval Finite Element Method with MATLAB*, Academic Press, San Diego, California.

[31] Nayak, S. (2020). Uncertain quantification of field variables in transient convection diffusion problems for imprecisely defined parameters. *International Communications in Heat and Mass Transfer*, 119, 104894.

https://doi.org/10.1142/9789811245367_0011

Chapter 11

Mathematical Modeling of Radon-Transport Mechanism with Imprecise Parameters

T. D. Rao*,‡ and S. Chakraverty†,§

Department of Mathematics,
Amrita School of Engineering,
Amrita Vishwa Vidyapeetham, Chennai, India
†*Department of Mathematics,*
National Institute of Technology Rourkela,
Odisha 769008, India
‡*dillu2.ou@gmail.com*
§*sne_chak@yahoo.com*

Abstract

In this chapter, the anomalous behavior of the radon data generated in a soil chamber is investigated by estimating an interval band using an interval-midpoint approach. Further, the mathematical formulation of radon transport from soil to buildings in an uncertain environment is also investigated. For the validation of the obtained results, we have used the deterministic results of the considered model problems and found good agreement between the results.

Keywords: finite difference, parameters, classical, imprecise, uncertainty, interval

1. Introduction

Radon is well known as a radioactive inert gas, which is usually found in rocks, soils, and groundwater. It was reported that 60.4 percent [1] of total indoor radon is originating from the underground. And also, radon has been monitored as an earthquake precursor for decades, along with various other precursors. However, the prediction of earthquakes based on these precursors is still elusive. One of the main reasons for this is the uncertainty involved in the estimated values of the precursors. The correct interpretation of the involved parameters of radon-transport mechanism is significant in monitoring the radon concentration. There exist various physical factors *viz.* radon concentration, advection, and diffusion coefficients, which affect the radon generation, and these physical factors are usually estimated by experiments. As such, while doing the experiments, one may have imprecise bounds or values of the effective parameters rather than exact values. So, handling radon-transport mechanism by considering parameters as imprecise is a challenging task.

Many researchers worked on the radon-transport mechanism by considering the involved parameters as deterministic or exact. As such, two methods for measuring 222 Rn exhalation and 226 Ra in soil samples were studied by Escobar *et al.* [2]. Kozak *et al.* [3] described modeling aspects of radium and radon transports through soil and vegetation. The release of radon isotopes under conditions of combined diffusion and flow from a fractured, semi-infinite medium, such as soil, is analyzed by Schery *et al.* [4]. Van der Spoel *et al.* [5] proposed a model to characterize the estimations of diffusive and advective radon transports in soil under well-defined and controlled conditions. The mechanism of radon generated within a few meters from the top of the Earth's surface by radioactive decay and description of the diffusion and advection parameters have been investigated by Nazaroff [6]. Kohl *et al.* [7] have investigated the numerical modeling of radon-transport mechanisms from subsurface soil to buildings with explicit Finite-Difference Method (EFDM). Modeling radon-transport equation, which describes the flow of radon from cracks in soil and concrete slabs, by using numerical methods has been investigated by Dimbylow *et al.* [8]. Savovic *et al.* [9] discussed the resources and processes which affect the radon transport from subsurface soil into buildings. Radon-transport mechanisms are modeled by governing differential equations, and such Differential Equations (DEs) are investigated by analytical or numerical methods [10, 11]. In this regard, some literature related to numerical modeling of various DEs is described here. A systematic introduction to Partial-Differential Equations (PDEs) and Finite-Element Method (FEM) are given by Solin [12]. Smith [13] investigated the stability, consistency, and convergence of hyperbolic, parabolic, and elliptic equations modeled by standard FDM. A modified-equation approach to analyze the stability and accuracy of FDMs have been proposed by Warming and Hyett [14]. Crank and Nicolson [15] proposed a method for evaluating a numerical solution of the nonlinear PDEs of the type which arises in problems of heat flow. An effective numerical method for solving PDEs in problems when traditional Eulerian and Lagrangian techniques fail was described by Braun and Sambridge [16].

Further, as described above, the involved coefficients and variables of differential equations are considered usually as deterministic or exact values, but while doing experiments, there may be a chance of impreciseness in observations and calculations. As such, one may have incomplete data about the coefficients and parameters. Such imprecise data may be considered as intervals or fuzzy numbers. In this regard, modeling differential equations with fuzzy/interval parameters is a challenging task. As such, Stefanini and Bede [17] proposed Hukuhara differentiability of interval-differential equations. An interval-difference approach for modeling the Poisson equation by using the conventional central difference has been discussed by Hoffmann and Marciniak [18]. A new interval midpoint technique to solve nth-order linear interval-differential equations with uncertain initial conditions has been proposed by Tapaswini and Chakraverty [19]. Nickel [20] described interval methods for the numerical solution of Ordinary-Differential Equations (ODEs).

As such, the present section gives the introduction related to the problem. The interval-midpoint approach is discussed in Section 2. In Section 3, an interval band to radon transport in a soil chamber is given. The radon-transport mechanism with imprecise parameters is discussed in Section 4. Section 5 includes numerical results and discussion. Finally, conclusions are drawn in the last section.

2. Interval-Midpoint Approach

A new midpoint approach is used to solve nth-order linear PDEs with interval uncertainty [23]. As such, to know the basic concepts of interval numbers one may refer Alefeld and Herzberger [21] and Moore *et al.* [22].

Let us consider the nth-order linear PDE with interval uncertainty in general form as

$$a_m(x)\frac{\partial^m \widetilde{z}}{\partial x^m} + a_{m-1}(x)\frac{\partial^{m-1}\widetilde{z}}{\partial x^{m-1}} + a_{m-2}(x)\frac{\partial^{m-2}\widetilde{z}}{\partial x^{m-2}} + \cdots$$

$$+ a_1(x)\frac{\partial \widetilde{z}}{\partial x} + a_0(x)\widetilde{z} = \widetilde{g}(x), \tag{1}$$

subject to initial conditions with uncertainty as

$$\widetilde{z}(0) = \widetilde{b}_0, \quad \frac{\partial \widetilde{z}(0)}{\partial x} = \widetilde{b}_1, \ldots, \quad \frac{\partial^{m-1}\widetilde{z}(0)}{\partial x^{m-1}} = \widetilde{b}_{m-1},$$

where $a_i(x), 0 \le i \le m$ can also be taken as interval values along with $b_i(x), 0 \le i \le m-1$.

By using the definition of Hukuhara derivative [17], we may write the PDE in Eq. (1) as

$$a_m(x)\left[\frac{\partial^m \underline{z}}{\partial x^m}, \frac{\partial^m \overline{z}}{\partial x^m}\right] + a_{m-1}(x)\left[\frac{\partial^{m-1}\underline{z}}{\partial x^{m-1}}, \frac{\partial^{m-1}\overline{z}}{\partial x^{m-1}}\right] + \cdots$$

$$+ a_1\left[\frac{\partial \underline{z}}{\partial x}, \frac{\partial \overline{z}}{\partial x}\right] + a_0[\underline{z}, \overline{z}] = [\underline{g}(x), \overline{g}(x)], \tag{2}$$

subject to initial conditions (with uncertainty) as

$$\left[\frac{\partial^{m-1}\underline{z}(0)}{\partial x^{m-1}}, \frac{\partial^{m-1}\overline{z}(0)}{\partial x^{m-1}}\right] = [\underline{b}_{m-1}, \overline{b}_{m-1}],$$

$$\left[\frac{\partial^{m-2}\underline{z}(0)}{\partial x^{m-2}}, \frac{\partial^{m-2}\overline{z}(0)}{\partial x^{m-2}}\right] = [\underline{b}_{m-2}, \overline{b}_{m-2}], \ldots,$$

$$\left[\frac{\partial \underline{z}(0)}{\partial x}, \frac{\partial \overline{z}(0)}{\partial x}\right] = [\underline{b}_1, \overline{b}_1], \ [\underline{z}(0), \overline{z}(0)] = [b_0, b_0].$$

In general, three cases arise with respect to the sign of the coefficients. As such, we have (1) coefficients $a_{m-1}(x), a_{m-2}(x), \ldots, a_0(x)$ all positive, (2) coefficients $a_{m-1}(x), a_{m-2}(x), \ldots, a_0(x)$ all negative, and

(3) both positive and negative coefficients. For simple understanding here, we are considering the coefficients $a_{m-1}(x), a_{m-2}(x), \ldots, a_{m-l}(x)$ are positive and $a_{m-l-1}(x)$, $a_{m-l-2}(x), \ldots, a_0(x)$ are negative.

Solving procedure of case 1:
If all the involved coefficients of Eq. (2) are positive, then the mid-point equation of Eq. (2) may be written as

$$a_m(x)\frac{\partial^m z_c}{\partial x^m} + a_{m-1}(x)\frac{\partial^{m-1} z_c}{\partial x^{m-1}} + a_{m-2}(x)\frac{\partial^{m-2} z_c}{\partial x^{m-2}} + \cdots$$

$$+ a_1(x)\frac{\partial z_c}{\partial x} + a_0(x) z_c = g_c(x), \tag{3}$$

with respect to the initial conditions

$$z_c(0) = b_{c_0}, \quad \frac{\partial z_c(0)}{\partial x} = b_{c_1}, \quad \frac{\partial^2 z_c(0)}{\partial x^2} = b_{c_2}, \ldots, \quad \frac{\partial^{m-1} z_c(0)}{\partial x^{m-1}} = b_{c_{m-1}}.$$

The lower-bound equation from Eq. (2) is

$$a_m(x)\frac{\partial^m \underline{z}}{\partial x^m} + a_{m-1}(x)\frac{\partial^{m-1} \underline{z}}{\partial x^{m-1}} + \cdots + a_1\frac{\partial \underline{z}}{\partial x} + a_0\underline{z} = \underline{g}(x), \tag{4}$$

subject to lower-bound initial conditions

$$\frac{\partial^{m-1}\underline{z}(0)}{\partial x^{m-1}} = \underline{b}_{m-1}, \frac{\partial^{m-2}\underline{z}(0)}{\partial x^{m-2}} = \underline{b}_{m-2}, \ldots, \quad \frac{\partial \underline{z}(0)}{\partial x} = \underline{b}_1, \underline{z}(0) = b_0.$$

Similarly, one may obtain the upper-bound equation as

$$a_m(x)\frac{\partial^m \overline{z}}{\partial x^m} + a_{m-1}(x)\frac{\partial^{m-1} \overline{z}}{\partial x^{m-1}} + \cdots + a_1\frac{\partial \overline{z}}{\partial x} + a_0\overline{z} = \overline{g}(x), \tag{5}$$

subject to upper-bound initial conditions

$$\frac{\partial^{m-1}\overline{z}(0)}{\partial x^{m-1}} = \overline{b}_{m-1}, \frac{\partial^{m-2}\overline{z}(0)}{\partial x^{m-2}} = \overline{b}_{m-2}, \ldots, \quad \frac{\partial \overline{z}(0)}{\partial x} = \overline{b}_1, \overline{z}(0) = b_0.$$

Now, one can obtain the lower-, midpoint-, and upper-bound solutions by using any standard analytical/numerical method to solve Eqs. (3)–(5). Further, a similar procedure is applicable for case 2.

Interval-midpoint approach is a useful method to handle physical problems of the form in case 3. As such, here, we are presenting a detailed description of case 3.

From Eq. (2), the lower-bound equation can be written as

$$a_m(x)\frac{\partial^m \underline{z}}{\partial x^m} + a_{m-1}(x)\frac{\partial^{m-1}\underline{z}}{\partial x^{m-1}} + \cdots + a_{m-l}(x)\frac{\partial^{m-l}\underline{z}}{\partial x^{m-l}}$$
$$+ a_{m-l-1}(x)\frac{\partial^{m-l-1}\overline{z}}{\partial x^{m-l-1}} + \cdots + a_1\frac{\partial \overline{z}}{\partial x} + a_0\overline{z} = \underline{g}(x), \quad (6)$$

Now, by using the midpoint value of an interval number ($\overline{z} = 2z_c - \underline{z}$), Eq. (6) can be represented as

$$a_m(x)\frac{\partial^m \underline{z}}{\partial x^m} + a_{m-1}(x)\frac{\partial^{m-1}\underline{z}}{\partial x^{m-1}} + \cdots + a_{m-l}(x)\frac{\partial^{m-l}\underline{z}}{\partial x^{m-l}}$$
$$- \left(a_{m-l-1}(x)\frac{\partial^{m-l-1}\underline{z}}{\partial x^{m-l-1}} + \cdots + a_1\frac{\partial \underline{z}}{\partial x} + a_0\underline{z}\right)$$
$$+ 2\left[a_{m-l-1}(x)\frac{\partial^{m-l-1}z_c}{\partial x^{m-l-1}} + \cdots + a_1\frac{\partial z_c}{\partial x} + a_0 z_c\right] = \underline{g}(x). \quad (7)$$

As such, one can obtain the midpoint solution from Eq. (3) and the lower-bound solution by solving Eq. (6). In a similar fashion, one may handle the upper-bound equation from Eq. (2). The initial conditions for the upper case may also be handled accordingly.

For the practical applicability of the proposed method, the following sections present application problems related to radon-diffusion mechanisms in uncertain environment. As such, radon-transport mechanism in a soil chamber to estimate the interval bounds for investigating anomalous behavior of generated radon concentration is presented in the following section for case 1.

3. Interval Bounds to Radon Transport in Soil Chamber

As discussed, radon is one of the known earthquake precursors. So, predicting the behavior of radon data is an important task. In general, there will always be a certain uncertainty arising in the generated radon data. But due to some external effects (earthquake, tsunami, etc.) and climatic conditions, there exists anomalous behavior (a certain rise or fall) in the radon data. As such, identifying the peaks in the data to predict the anomalous behavior of data plays an important role in the prediction of external effects. In this regard, here, we are presenting modeling of radon transport in a soil chamber by the proposed method of giving interval bounds which are helpful in predicting the anomalous behavior of the data.

3.1. *Radon Buildup in a Soil Chamber*

In this study, an accumulation radon chamber experiment was conducted at NIT Rourkela to study the changes in radon exhalation from the Earth's surface. As such, a radon soil chamber (Figure 1) was placed on a radon source to build up radon inside the soil chamber [24]. From this chamber, the generated radon is transferred to the measuring device (radon monitor (Figure 2)) to estimate the radon concentration level at closed-interval time [25]. One of the best

Figure 1. Soil chamber.

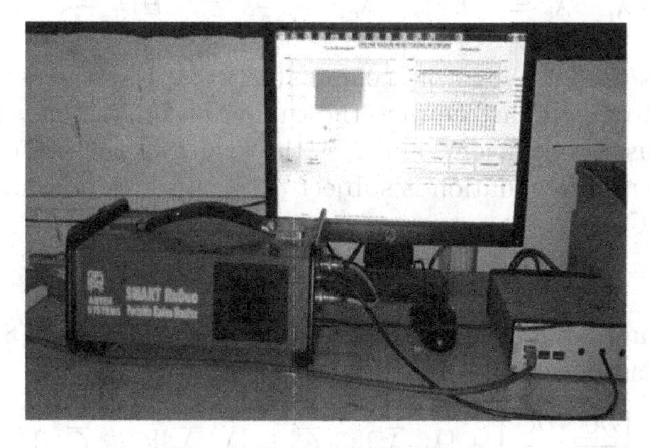

Figure 2. Radon monitor.

approaches for determining the radon transport from soil is the online radon-monitor technique. A soil chamber is a single unit whose open side is inserted into the soil. And the space above the soil surface bounded by the chamber is called the headspace of the soil chamber. In the present targeted work, we have taken the radon data which has anomalous behavior from the soil chamber experiment and formulated it to give interval bounds to the considered data.

3.2. *Mathematical Modeling by Interval-Midpoint Approach*

In general, the physical parameters on which radon generation depends, *viz.* radium concentration, porosity, and diffusion coefficients are usually measured experimentally. As such, while doing the soil-chamber experiment, the estimated values of the involved parameters may deviate significantly from the actual values. In this regard, estimating a general uncertainty band to radon transport in the soil chamber is an important task to predict the anomalous behavior of the generated data. The radon transport in a soil chamber with uncertain parameters may be presented by the transport equation as

$$\frac{\partial \widetilde{C}}{\partial t} + \widetilde{\lambda}_e \widetilde{C} = \widetilde{\lambda}_c \widetilde{C}_\infty, \tag{8}$$

where λ_e is the effective decay constant, λ_c is the diffusion-leak rate, and λ is the radon-decay constant (2.1×10^{-6}), and

$$\lambda_e = \lambda_c + \lambda, \quad \lambda_e = \frac{1}{t_1} h^{-1}, \quad \lambda_c = \frac{H}{DL} s^{-1}, \quad H = \frac{V}{A}, \quad L = \sqrt{\frac{D}{\lambda}},$$

where A is the base area of accumulator(πr^2), V is the accumulator volume above soil surface plus the tubing and detector volumes, D is the diffusion coefficient, and L is the length of the soil chamber.

The transport equation is subject to the uncertain boundary conditions (BCs),

$$\widetilde{C}(t = 0) = \widetilde{C}_0, \quad \widetilde{C}(t = \infty) = \widetilde{C}_\infty.$$

Now, by applying the Hukuhara derivative [17], Eq. (8) can be expressed as

$$\left[\frac{\partial \underline{C}}{\partial t}, \frac{\partial \overline{C}}{\partial t}\right] + [\underline{\lambda}_e, \overline{\lambda}_e][\underline{C}, \overline{C}] = [\underline{\lambda}_c, \overline{\lambda}_c][\underline{C}_\infty, \overline{C}_\infty], \tag{9}$$

subject to interval BCs,

$$[\underline{C},\overline{C}](t=o)=[\underline{C}_0,\overline{C}_0],\quad [\underline{C},\overline{C}](t=\infty)=[\underline{C}_\infty,\overline{C}_\infty].$$

Now, one can obtain the midpoint equation from Eq. (9) as

$$\frac{\partial C_c}{\partial t}+(\lambda_e)_c C_c=(\lambda_c)_c(C_\infty)_c, \tag{10}$$

with respect to the BCs

$$C_c(t=0)=(C_0)_c,\quad C_c(t=\infty)=(C_\infty)_c.$$

And the analytical solution of Eq. (10) may be obtained as

$$C_c(t)=(C_{0_c})-\frac{(\lambda_c)_c}{(\lambda_e)_c}(C_\infty)_c e^{-(\lambda_e)_c t}+\frac{(\lambda_c)_c}{(\lambda_e)_c}(C_\infty)_c. \tag{11}$$

In a similar fashion, one may obtain the following lower- and upper-bound model equations from Eq. (9):

$$\frac{\partial \underline{C}}{\partial t}+\underline{\lambda}_e\underline{C}=\underline{\lambda}_c\underline{C}_\infty \tag{12}$$

and

$$\frac{\partial \overline{C}}{\partial t}+\overline{\lambda}_e\overline{C}=\overline{\lambda}_c\overline{C}_\infty. \tag{13}$$

Further, one can solve the obtained bound equations using any standard method along with their respective boundary conditions to get interval bounds to the data generated in a soil-chamber experiment to predict the anomalous behavior.

3.3. *Experimental Results and Discussion*

Numerical estimations to find interval bounds of the physical parameters involved in the considered model are investigated here. As such, the exponential fittings to the considered radon data are used to estimate the involved interval parameters. Further, the comparisons of the results with the evaluated physical parametric values are depicted as figures in subsequent discussions.

Initially, we have considered experimental data which has an anomalous peak from the soil-chamber experiment. Then, an exponential fitting, $y(t)=A_1e^{-\frac{t}{t_1}}+y_0$, is used to estimate the involved

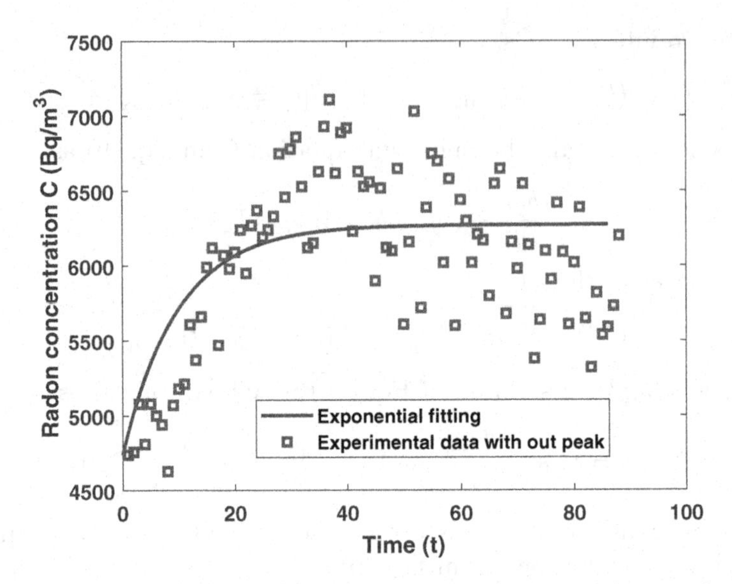

Figure 3. Exponential fitting to experimental data without peak.

Table 1. Exponential fitting values.

Variable	Numerical Value
y_0	6232.05239
A_1	−1875.1458
t_1	10.15878

parametric values of the midpoint model in Eq. (10). Figure 3 represents the exponential fitting of the soil-chamber experimental data without the anomalous peak, and Table 1 depicts the obtained fitted parametric details.

Then, by using the midpoint-estimated parametric details, we have estimated the lower- and upper-bound parametric details as presented in Table 2, and these parametric details are used in the obtained lower- and upper-bound model equations (Eqs. (12) and (13)). Further, we have estimated an interval band to the considered anomalous data. These interval bounds indicate the normal behavior of the considered radon data. As such, the estimation of interval bounds to the radon data plays an important role in the investigation of the behavior of the radon gas at the time of anomalous effects, such as earthquakes and tsunamis. Figure 4 presents the interval

Table 2. Estimated lower- and upper-bound parametric details by using the midpoint values.

Parametric Bounds	Estimated Bound Values
$[\underline{C}_0, \overline{C}_0]$	[4730, 4750]
$[\underline{D}, \overline{D}]$	[0.009872, 0.011995]
$[\underline{C}_\infty, \overline{C}_\infty]$	[5859.5166, 6614.1566]
$[\overline{\lambda}_e, \underline{\lambda}_e]$	[0.084173, 0.113494]
$[\overline{\lambda}_c, \underline{\lambda}_c]$	[0.084168, 0.113495]

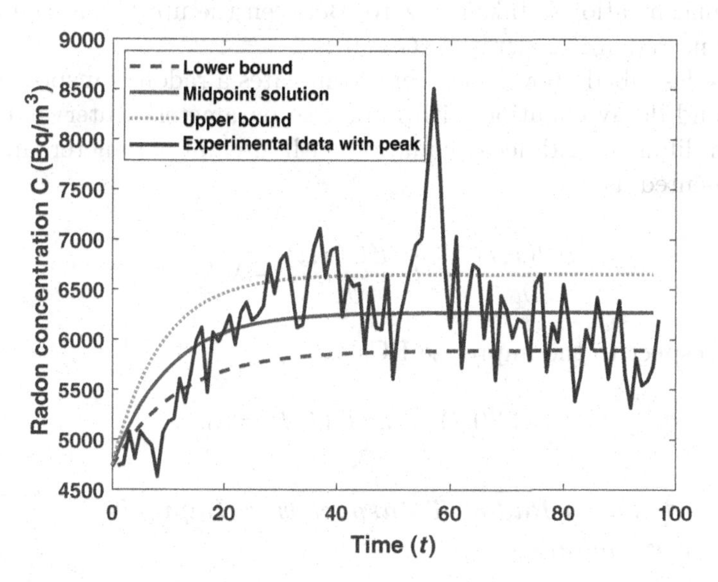

Figure 4. Interval bounds of the experimental data with a peak.

band of the considered anomalously behaved radon data along with the midpoint solution. In this regard, the midpoint approach plays an important role in the estimation of interval band to the anomalous data.

4. Radon-Transport Mechanism with Imprecise Parameters

The physical mechanisms which are responsible for the transport of radon from subsurface to buildings were described by Nazaroff [6].

A general transport equation which describes diffusion and decay of radon transport in soil pore air can be defined by the transport equation (Savovic *et al.* [9]) as

$$\frac{\partial C(x,t)}{\partial t} = D\frac{\partial^2 C(x,t)}{\partial x^2} - \lambda C(x,t) + G, \tag{14}$$

where D is the diffusion coefficient ($m^2 s^{-1}$), λ is the decay constant (s^{-1}), and G is the radon-generation rate. At $x = L$ (lower end), the radon concentration is assumed as c_0, and at surface ($x = 0$), the concentration is taken as zero. Between the upper and the lower ends, no radon is generated ($G = 0$).

As described above, this work formulates a radon transport diffusion and decay equation with imprecise parameters (intervals). The radon diffusion and decay equation with interval parameters may be represented as

$$\frac{\partial \widetilde{C}(x,t)}{\partial t} = \widetilde{D}\frac{\partial^2 \widetilde{C}(x,t)}{\partial x^2} - \lambda \widetilde{C}(x,t), \tag{15}$$

with respect to the imprecise BCs as

$$\widetilde{C}(0,t) = 0; \quad \widetilde{C}(L,t) = \widetilde{c}_0. \tag{16}$$

4.1. *Modeling Radon Transport with Imprecise Parameters*

From the said differential equation in Eq. (15), radon transport through soil with the imprecise parameters D and c_0 lead to the following three cases.

Case 1: When $D = [\underline{D}, \overline{D}]$ is the interval number, and c_0 is an exact or deterministic value.

Case 2: $c_0 = [\underline{c_0}, \overline{c_0}]$ is the interval, and D is an exact or deterministic value.

Case 3: Both D and c_0 are interval numbers.

Case 1: D is considered as the interval and c_0 is taken as an exact or deterministic value. Then, the said differential equation in Eq. (15)

can be represented as

$$\frac{\partial \widetilde{C}(x,t)}{\partial t} = \widetilde{D}\frac{\partial^2 \widetilde{C}(x,t)}{\partial x^2} - \lambda \widetilde{C}(x,t), \tag{17}$$

with respect to the BCs,

$$C(0,t) = 0, \quad C(L,t) = c_0. \tag{18}$$

It may be noted that c_0 is an exact or deterministic value here. The above uncertain differential equation (Eq. (17)) can be represented as follows (using Hukuhara derivative):

$$\left[\frac{\partial \underline{C}(x,t)}{\partial t}, \frac{\partial \overline{C}(x,t)}{\partial t}\right] = [\underline{D}, \overline{D}]\left[\frac{\partial^2 \underline{C}(x,t)}{\partial x^2}, \frac{\partial^2 \overline{C}(x,t)}{\partial x^2}\right] - \lambda[\underline{C}(x,t), \overline{C}(x,t)], \tag{19}$$

subject to the BCs,

$$[\underline{C}(0,t), \overline{C}(0,t)] = [0,0], \quad [\underline{C}(L,t), \overline{C}(L,t)] = [c_0, c_0]. \tag{20}$$

From Eq. (19), the lower bound of radon-transport equation may be symbolized as

$$\frac{\partial \underline{C}(x,t)}{\partial t} = \underline{D}\frac{\partial^2 \underline{C}(x,t)}{\partial x^2} - \lambda \overline{C}(x,t), \tag{21}$$

with respect to the BCs,

$$\underline{C}(0,t) = 0, \quad \underline{C}(L,t) = c_0. \tag{22}$$

By using the known center value of interval, we can represent the lower-bound radon-transport equation in Eq. (21) as

$$\frac{\partial \underline{C}(x,t)}{\partial t} = \underline{D}\frac{\partial^2 \underline{C}(x,t)}{\partial x^2} + \lambda \underline{C}(x,t) - 2\lambda C_c(x,t). \tag{23}$$

The EFDM has been used (Savovic *et al.* [9] and Ames [11]) to solve the transport equation (Eq. (23)). The central-difference scheme was used to represent the term $\frac{\partial^2 \underline{C}(x,t)}{\partial x^2} = (\underline{C}_{i+1,j} - 2\underline{C}_{i,j} + \underline{C}_{i-1,j})/\Delta x^2$

and for $\frac{\partial \underline{C}(x,t)}{\partial t} = (\underline{C}_{i,j+1} - \underline{C}_{i,j})/\Delta t$ (the forward scheme is used). With these substitutions, Eq. (23) transforms into (Solin [12] and Smith [13])

$$\underline{C}_{i,j+1} = m_1\underline{C}_{i+1,j} + n_1\underline{C}_{i,j} + r_1\underline{C}_{i-1,j} - 2\lambda\Delta t C_{c,i,j}, \qquad (24)$$

where $m_1 = \frac{\underline{D}\Delta t}{\Delta x^2}$, $n_1 = 1 + \lambda\Delta t - 2 \times \frac{\underline{D}\Delta t}{\Delta x^2}$, $r_1 = \frac{\underline{D}\Delta t}{\Delta x^2}$.

The corresponding lower and center BCs may be symbolized as

$$\underline{C}_{0,j} = 0, \qquad \underline{C}_{N,j}, = c_0, \qquad (25)$$

$$C_{c,0,j} = 0, \quad C_{c,N,j}, = c_0. \qquad (26)$$

The upper bound of radon-transport equation may also be obtained as

$$\overline{C}_{i,j+1} = m_2\overline{C}_{i+1,j} + n_2\overline{C}_{i,j} + r_2\overline{C}_{i-1,j} - 2\lambda\Delta t C_{c,i,j}, \qquad (27)$$

where $m_2 = \frac{\overline{D}\Delta t}{\Delta x^2}$, $n_2 = 1 + \lambda\Delta t - 2 \times \frac{\overline{D}\Delta t}{\Delta x^2}$, $r_2 = \frac{\overline{D}\Delta t}{\Delta x^2}$, with respect to upper and center BCs as

$$\overline{C}_{0,j} = 0, \qquad \overline{C}_{N,j}, = c_0, \qquad (28)$$

$$C_{c,0,j} = 0, \quad C_{c,N,j}, = c_0. \qquad (29)$$

Case 2: In this case, the uncertain differential equation (Eq. (15)) can be represented as follows (Hukuhara derivative):

$$\left[\frac{\partial \underline{C}(x,t)}{\partial t}, \frac{\partial \overline{C}(x,t)}{\partial t}\right] = D\left[\frac{\partial^2 \underline{C}(x,t)}{\partial x^2}, \frac{\partial^2 \overline{C}(x,t)}{\partial x^2}\right] - \lambda[\underline{C}(x,t), \overline{C}(x,t)], \qquad (30)$$

subject to the BCs,

$$[\underline{C}(0,t), \overline{C}(0,t)] = [0,0], [\underline{C}(L,t), \overline{C}(L,t)] = [\underline{c}_0, \overline{c}_0]. \qquad (31)$$

From Eq. (30), the lower bound of radon-transport equation can be represented as follows:

$$\frac{\partial \underline{C}(x,t)}{\partial t} = D\frac{\partial^2 \underline{C}(x,t)}{\partial x^2} - \lambda\overline{C}(x,t). \qquad (32)$$

By using the mid point of an interval, we can present the lower-bound radon-transport equation (Eq. (32)) as

$$\frac{\partial \underline{C}(x,t)}{\partial t} = D\frac{\partial^2 \underline{C}(x,t)}{\partial x^2} + \lambda\underline{C}(x,t) - 2\lambda C_c(x,t). \qquad (33)$$

By applying EFDM for this case, we may write as (Smith [13])

$$\underline{C}_{i,j+1} = m_3\underline{C}_{i+1,j} + n_3\underline{C}_{i,j} + r_3\underline{C}_{i-1,j} - C_{c,i,j}(2\lambda\Delta t), \tag{34}$$

where $m_3 = \frac{D\Delta t}{\Delta x^2}$, $n_3 = 1 + \lambda\Delta t - 2 \times \frac{D\Delta t}{\Delta x^2}$, $r_3 = \frac{D\Delta t}{\Delta x^2}$, with respect to its lower and center BCs as

$$\underline{C}_{0,j} = 0, \qquad \underline{C}_{N,j}, = \underline{c}_0, \tag{35}$$

$$C_{c,0,j} = C_{c,0}, \quad C_{c,N,j}, = 0. \tag{36}$$

Similarly, by applying the EFDM model for upper bound, one may obtain

$$\overline{C}_{i,j+1} = m_4\overline{C}_{i+1,j} + n_4\overline{C}_{i,j} + r_4\overline{C}_{i-1,j} - C_{c,i,j}(2\lambda\Delta t), \tag{37}$$

where $m_4 = \frac{\overline{D}\Delta t}{\Delta x^2}$, $n_4 = 1 + \lambda\Delta t - 2 \times \frac{\overline{D}\Delta t}{\Delta x^2}$, $r_4 = \frac{\overline{D}\Delta t}{\Delta x^2}$, with respect to lower and center BCs as

$$\overline{C}_{0,j} = 0, \qquad \overline{C}_{N,j}, = \overline{C}_0, \tag{38}$$

$$C_{c,0,j} = 0, \qquad C_{c,N,j}, = C_{c,0}. \tag{39}$$

Case 3: D and c_0 are both taken as intervals.

In this case, the lower bound of the radon-transport equation can be represented as

$$\frac{\partial \underline{C}(x,t)}{\partial t} = \underline{D}\frac{\partial^2 \underline{C}(x,t)}{\partial x^2} - \overline{v}\frac{\partial \overline{C}(x,t)}{\partial x} - \lambda\overline{C}(x,t). \tag{40}$$

In a similar fashion, by using the center value of $\widetilde{C}$, we can represent the lower-bound radon-transport equation (Eq. (40)) as follows:

$$\frac{\partial \underline{C}(x,t)}{\partial t} = \underline{D}\frac{\partial^2 \underline{C}(x,t)}{\partial x^2} + \lambda\underline{C}(x,t) - 2\lambda C_c(x,t). \tag{41}$$

Again, utilizing the EFDM for this case, we may have

$$\underline{C}_{i,j+1} = m_5\underline{C}_{i+1,j} + n_5\underline{C}_{i,j} + r_5\underline{C}_{i-1,j} - C_{c,i,j}(2\lambda\Delta t), \tag{42}$$

where $m_5 = \frac{\underline{D}\Delta t}{\Delta x^2}$, $n_5 = 1 + \lambda\Delta t - 2 \times \frac{\underline{D}\Delta t}{\Delta x^2}$, $r_5 = \frac{\underline{D}\Delta t}{\Delta x^2}$, with respect to lower and center BCs as

$$\underline{C}_{0,j} = 0, \quad \underline{C}_{N,j}, = \underline{c}_0, \tag{43}$$

$$C_{c,0,j} = 0, \quad C_{c,N,j} = c_{c,0}. \tag{44}$$

Further, applying the EFDM model for upper bound of radon-transport equation, we get

$$\overline{C}_{i,j+1} = m_6\overline{C}_{i+1,j} + n_6\overline{C}_{i,j} + r_6\overline{C}_{i-1,j} - C_{c,i,j}(2\lambda\Delta t), \tag{45}$$

where $m_6 = \frac{\overline{D}\Delta t}{\Delta x^2}$, $n_6 = 1 + \lambda\Delta t - 2 \times \frac{\overline{D}\Delta t}{\Delta x^2}$, $r_6 = \frac{\overline{D}\Delta t}{\Delta x^2}$, with respect to lower and center BCs as

$$\overline{C}_{0,j} = 0, \qquad \overline{C}_{N,j} = \overline{C}_0, \tag{46}$$

$$C_{c,0,j} = C_{c,0}, \quad C_{c,N,j} = 0. \tag{47}$$

5. Numerical Results and Discussion

A 10 m-deep domain was assumed in the vertical direction (Savovic *et al.* [9]). One side of the soil slab ($x = L$) was exposed to a constant high-radon concentration of $c_0 = 10^4 Bq/m^3$, while the concentration at the surface $C(x = 0) = 0$, and the model parameter (diffusion coefficient) has been taken as $D = 1.0 \times 10^{-6} m^2 s^{-1}$. Correspondingly, the uncertain model interval parameters have been considered as $\widetilde{D} = [\underline{D}, \overline{D}] = [0.9 \times 10^{-6}, 1.1 \times 10^{-6}] m^2 s^{-1}$ and concentration at $x = L$ as $\widetilde{c_0} = [\underline{c}_0, \overline{c}_0] = [9{,}500, 10{,}500] Bq/m^3$. Then, Eq. (15) has been solved by the proposed method by using the above-described parametric details. As such, the obtained midpoint solutions along with the steady-state solution of the considered problem are used for the comparison of the obtained results and good agreement was found.

As such, Figure 5 represents the radon concentration at different times, obtained by solving the midpoint equation of the considered problem (Eq. 15) (Savovic *et al.* [9]). Further, for all the considered three cases, we have estimated the interval bounds at different times. As such, Figures 6 and 7 represent interval bounds at different times $t = 10^5$ and $t = 10^6$, respectively, for case 1, and Figures 8 and 9 depict the interval bounds at different times $t = 10^5$ and $t = 10^6$, respectively, for case 2. In a similar fashion, Figures 10 and 11 give the upper-, center-, and lower-radon concentrations at different times $t = 10^5$ and $t = 10^6$, respectively, for case 3.

In this investigation, we have considered the same range of uncertainty for the parameters c_0 and D. Then, we have given interval

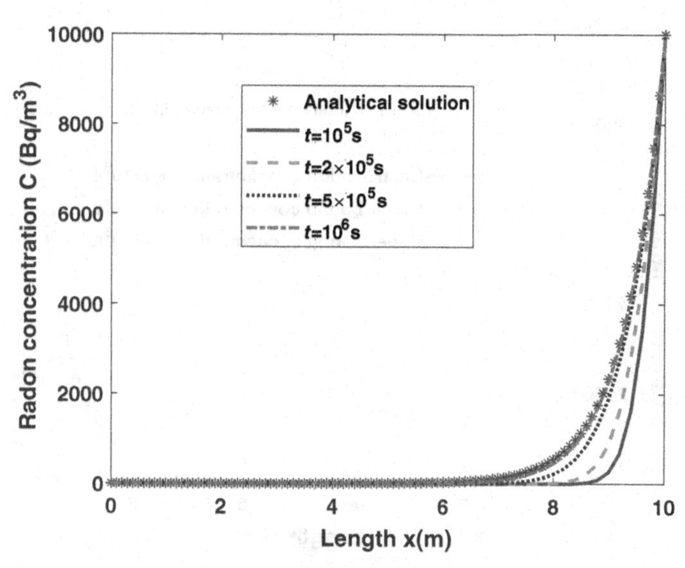

Figure 5. Radon concentration results using EFDM and analytical solution (steady state).

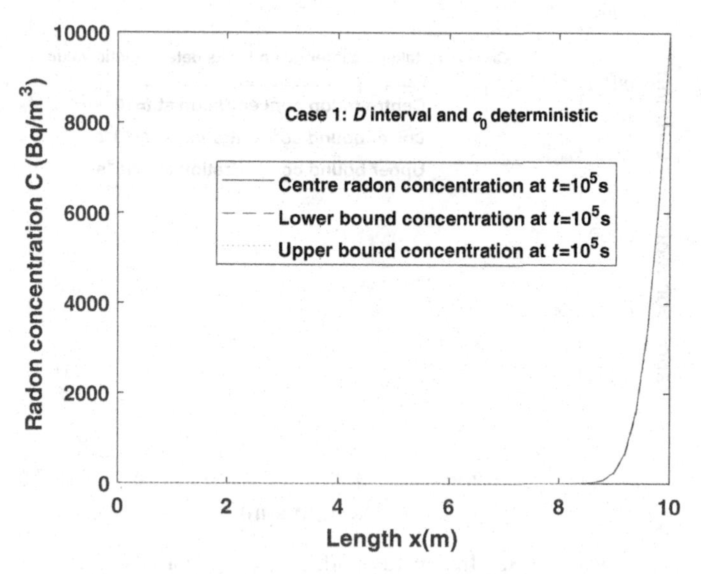

Figure 6. Interval bounds at $t = 10^5$ for case 1.

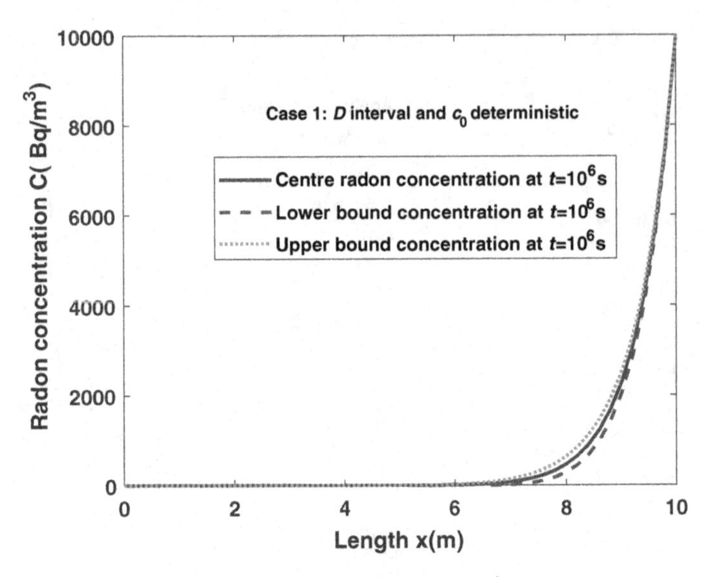

Figure 7. Interval bounds at $t = 10^6$ for case 1.

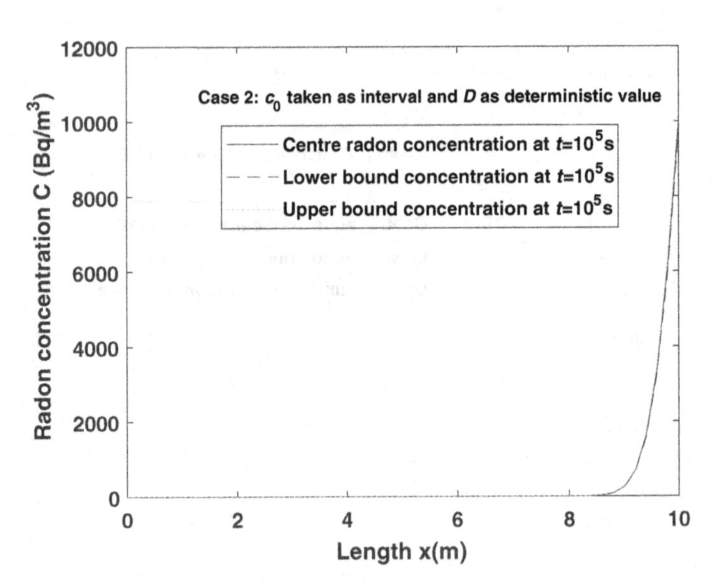

Figure 8. Interval bounds at $t = 10^5$ for case 2.

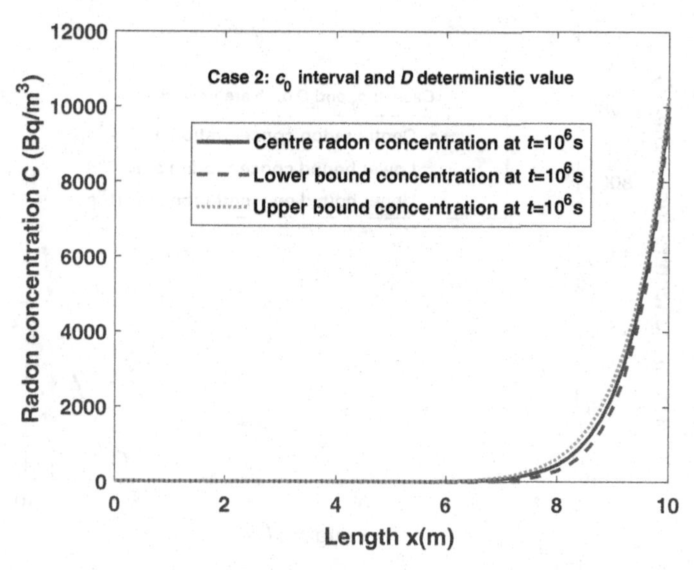

Figure 9. Interval bounds at $t = 10^6$ for case 2.

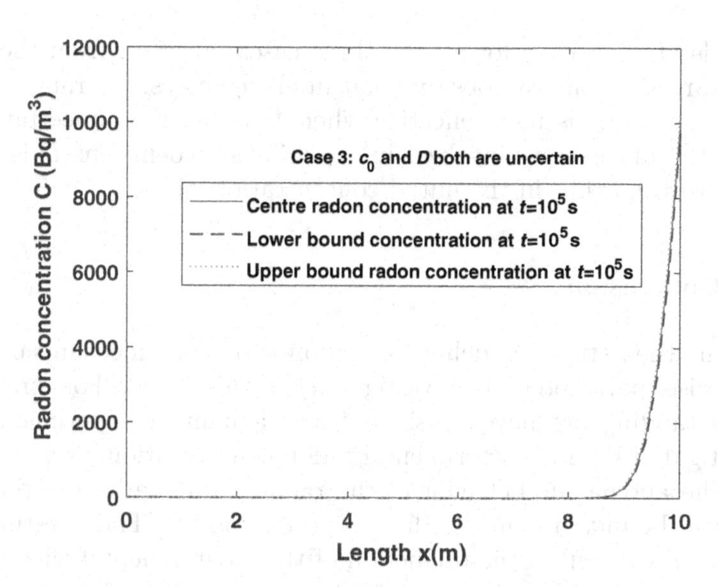

Figure 10. Interval bounds at $t = 10^5$ for case 3.

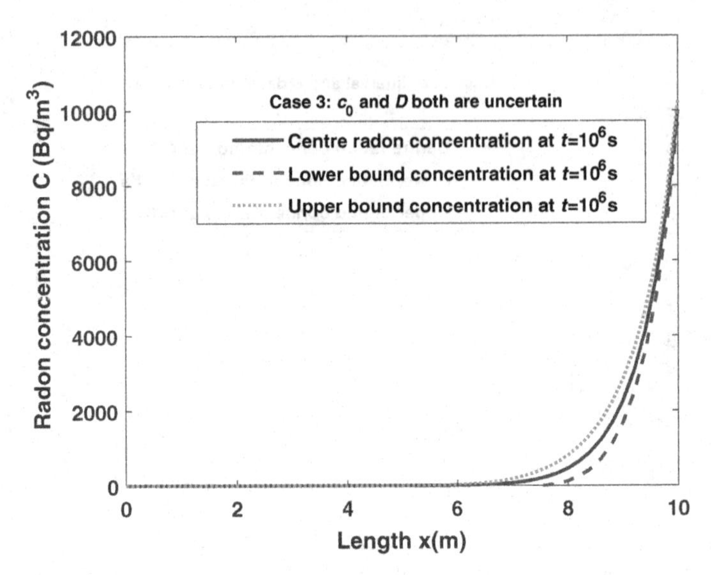

Figure 11. Interval bounds at $t = 10^6$ for case 3.

bounds at each time for all the three cases. Further, from the presented results, one can observe that in all the cases, the range of the interval bounds is more effective when D is taken as uncertain. As such, the uncertainty involved in the diffusion coefficient D is more effective than that in the initial concentration c_0.

6. Conclusion

The mathematical modeling of radon-transport mechanism with imprecise parameters is investigated in this work. For practical understanding, we have considered a soil-chamber experiment and investigated the real experimental radon-concentration data. To predict the anomalous behavior of the radon data, we have estimated interval bounds by using the proposed model. These estimated bounds are useful in the prediction of the environmental effects (climatic conditions, earthquakes, tsunamis, etc.) on radon concentrations. Further, we have considered radon-transport mechanism from subsurface soil to buildings in uncertain environments and applied the proposed method. Then, we have investigated the effects and importance of the imprecise parameters involved in radon-transport mechanism and given interval bounds at different times of the considered model. Finally, comparison of the results is done with the deterministic/classical results of the considered problems.

Acknowledgment

The authors are very much thankful to the Board of Research in Nuclear Sciences (BRNS), Mumbai, India for the support to the present research work. Also, we are extremely thankful to B. K. Sahoo and J. J. Gaware for their support and guidance during the experimental works.

References

[1] Ren, T. (2001). Source, level and control of indoor radon. *Radiation Protection (Taiyuan)*, 21(5), 291–299.
[2] Escobar, V., Gomez, F., Tome, V., and Lozano. (1999). Procedures for the determination of 222Rn exhalation and effective 226Ra activity in soil samples. *Applied Radiation and Isotopes*, 50(6), 1039–1047.
[3] Kozak, J. A., Reeves, H. W., and Lewis, B. A. (2003). Modeling radium and radon transport through soil and vegetation. *Journal of Contaminant Hydrology*, 66(3–4), 179–200.
[4] Schery, S. D., *et al.* (1988). The flow and diffusion of radon isotopes in fractured porous media: Part 2, Semi-infinite media. *Radiation Protection Dosimetry*, 24(1–4), 191–197.
[5] Van der Spoel, W. H., Van der Graaf, E. R., and De Meijer, R. J. (1998). Combined diffusive and advective transport of radon in a homogeneous column of dry sand. *Health Physics*, 74(1), 48–63.
[6] Nazaroff, W. (1992). William Radon transport from soil to air. *Reviews of Geophysics*, 30(2), 137–160.
[7] Kohl, T., Medici, F., and Rybach, L. (1994). Numerical simulation of radon transport from subsurface to buildings. *Journal of Applied Geophysics*, 31(1–4), 145–152.
[8] Dimbylow, P. J. and Wilkinson, P. (1985). The numerical solution of the diffusion equation describing the flow of radon through cracks in a concrete slab. *Radiation Protection Dosimetry*, 11(4), 229–236.

[9] Savović, S., Djordjevich, A., and Ristić, G. (2011). Numerical solution of the transport equation describing the radon transport from subsurface soil to buildings. *Radiation Protection Dosimetry*, 150(2), 213–216.
[10] Ames, W. F. (2014). *Numerical Methods for Partial Differential Equations*, Academic press.
[11] Langtangen, H. P. (1998). Computational partial differential equations — numerical methods and diffpack programming. Lecture notes in computational science and engineering.
[12] Solín, P. (2005). *Partial Differential Equations and the Finite Element Method*, Vol. 73, John Wiley and Sons.
[13] Smith, G. D. (1985). *Numerical Solution of Partial Differential Equations: Finite Difference Methods*, Oxford University Press.
[14] Warming, R. F. and Hyett, B. J. (1974). The modified equation approach to the stability and accuracy analysis of finite-difference methods. *Journal of Computational Physics*, 14(2), 159–179.
[15] Crank, J. and Nicolson, P. (1947). A practical method for numerical evaluation of solutions of partial differential equations of the heat-conduction type. *Mathematical Proceedings of the Cambridge Philosophical Society*, 43(1), 50–67. Cambridge University Press.
[16] Braun, J. and Sambridge, M. (1995). A numerical method for solving partial differential equations on highly irregular evolving grids. *Nature*, 376(6542), 655.
[17] Stefanini, L. and Bede, B. (2009). Generalized Hukuhara differentiability of interval-valued functions and interval differential equations. *Nonlinear Analysis: Theory, Methods and Applications*, 71(3), 1311–1328.
[18] Hoffmann, T. and Marciniak, A. (2013). Solving the Poisson equation by an interval difference method of the second order. *Computational Methods in Science and Technology*, 19(1), 13–21.
[19] Tapaswini, S. and Chakraverty, S. (2014). New midpoint-based approach for the solution of n-th order interval differential equations. *Reliable Computing*, 20, 25–44.
[20] Nickel, K. L. (1986). Using interval methods for the numerical solution of ODE's. *ZAMM — Journal of Applied Mathematics and Mechanics, Zeitschrift fur Angewandte Mathematik und Mechanik*, 66(11), 513–523.
[21] Alefeld, G. and Herzberger, J. (1983). *Introduction to Interval Computations*, Academic Press: New York.
[22] Moore, R. E., Kearfott, R. B., and Cloud, M. J. (2009). Introduction to interval analysis. *Society for Industrial and Applied Mathematics*.

[23] Chakraverty, S., Tapaswini, S., and Behera, D. (2016). *Fuzzy Differential Equations and Applications for Engineers and Scientists*, CRC Press Taylor and Francis Group: Boca Raton, USA.
[24] Menetrez, M. Y., Mosley, R. B., Snoddy, R., and Brubaker Jr., S. A. (1996). Evaluation of radon emanation from soil with varying moisture content in a soil chamber. *Environment International*, 22, 447–453.
[25] Sahoo, B. K. and Mayya, Y. S. (2010). Two dimensional diffusion theory of trace gas emission into soil chambers for flux measurements. *Agricultural and Forest Meteorology*, 150(9), 1211–1224.

https://doi.org/10.1142/9789811245367_0012

Chapter 12

An Efficient Numerical Scheme for Time-Fractional Coupled Shallow-Water Equations with a Non-Singular Fractional Derivative

Amit Kumar
Department of Mathematics, Balarampur College, Purulia, West Bengal, India
amitnit.jsr123@gmail.com

Abstract

In this chapter, an analytical technique, namely Modified Homotopy Analysis Transform Method (MHATM), is extended to solve time-fractional coupled Shallow-Water Equations (SWEs). SWEs are a system of partial-differential equations that illustrate the flow below a pressure surface in a fluid. The novelty of the proposed algorithm is that an exponential form of the non-singular kernel of the fractional derivative, known as Caputo–Fabrizio (CF) derivative, is taken. With the help of Banach's fixed-point theory, the uniqueness and convergence analysis for the coupled SWEs are presented through the theorems. The MHATM with CF fractional derivative constructs the solution in the form of a convergent series in a large acceptable region, which is helpful to adjust the region of convergence of a series solution. The obtained results have been exposed through different graphical analysis with the variation of the diverse values of the fractional parameter. However, the outputs indicate that the mentioned algorithm provides accuracy and is applicable to solve various nonlinear, coupled fractional problems, such as SWEs.

Keywords: Homotopy analysis transform method, shallow-water equations, Caputo–Fabrizio derivative, uniqueness and convergence analysis, homotopy polynomial

1. Introduction

In our day-to-day life, there are several physical phenomena which are illustrated by the set of nonlinear partial-differential equations. Such type of equations arise in various areas of mathematics, physics, fluid dynamics, and chemistry [1, 2]. In the case of fluid dynamics, the nonlinear equations are appearing in the circumstances of shallow-water equations (SWEs). The SWEs are a system of nonlinear partial-differential equations that occur in a fluid flow. The most important properties of the SWEs are that the fluid is homogeneous and incompressible, the pressure distribution is hydrostatic, and the flow is steady. Other uses of SWEs are in tidal flats, ocean currents, coastal regions, and to study dredging feasibility [3, 4].

In the present research, the application of fractional calculus is gaining popularity among researchers. The reason is that it has enormous applicability to real-world problems [5, 6]. Fractional-Differential Equations (FDEs) are the most general edition of integer-order differential equations. In the existing literature, there are several fractional derivatives which have singular kernels, for instance, Riemann–Liouville derivative, Hadamard derivative, Miller–Ross derivative, Marchand derivative, Riesz derivative, and Caputo derivative [7–9]. Based on the exponential form of the non-singular kernel, eminent researchers Caputo and Fabrizio proposed a fractional derivative, namely Caputo–Fabrizio (CF) fractional derivative [10]. Afterward, this derivative was successfully applied to solve different fractional problems [11, 12]. The various theories and applications of CF as well as the methods for finding the solutions of FDEs are explained in [13–16].

In this paper, we have taken the following form of the time-fractional SWEs [17, 18]:

$$\begin{cases} {}^{CF}D_t^{\rho}\Psi(\xi,t) + \Phi(\xi,t)\Psi_{\xi}(\xi,t) + \Psi(\xi,t)\Phi_{\xi}(\xi,t) = 0, \\ {}^{CF}D_t^{\delta}\Phi(\xi,t) + \Phi(\xi,t)\Phi_{\xi}(\xi,t) + \Psi_{\xi}(\xi,t) = 0, \end{cases} \tag{1}$$

where $\Psi(\xi,t)$ indicates the free surface and $\Phi(\xi,t)$ indicates the horizontal velocity component of the fluid flow. $0<\rho\leq 1$ and $0<\delta\leq 1$ are the fractional derivatives taken in CF sense [10]. We have used Modified Homotopy Analysis Transform Method (MHATM) to find out the approximate analytical solutions of Eq. (1). MHATM is a graceful mixture of the homotopy-analysis method, Laplace transform method, and the homotopy polynomial [19]. MHATM is very much used by many researchers for finding the solutions of different types of fractional problems [20–22].

The rest of the paper is organized as follows: In Section 2, the fundamental concepts of the CF derivative is illustrated. Implementation of the MHATM to time-fractional SWEs is discussed in Section 3. In Section 4, the convergence analysis of the MHATM solution of SWEs is presented. Numerical results and discussions are presented in Section 5. Finally, we close up with concluding remarks in Section 6.

2. Preliminaries

This part discusses some basic definitions related to CF derivatives, which are used in the current work.

Definition 1. Supposing that $\Psi(\xi,t)$ is a function such that $\Psi\in K^1(a,b)$, where $a<b$. If $0<\rho\leq 1$, then the CF fractional derivative is defined as [10–12]

$${}^{CF}D_t^{\rho}\Psi(\xi,t) = \frac{B(\rho)}{(1-\rho)}\int_a^t \Psi'(\xi,\tau)e^{-\frac{\rho(t-\tau)}{1-\rho}}\,d\tau. \tag{2}$$

By putting $a=0$, we have

$${}^{CF}D_t^{\rho}\Psi(\xi,t) = \frac{B(\rho)}{(1-\rho)}\int_0^t \Psi'(\xi,\tau)e^{-\frac{\rho(t-\tau)}{1-\rho}}\,d\tau, \tag{3}$$

where $B(\rho)$ is a normalization function which depends on ρ, satisfying $B(0) = B(1) = 1$.

Remarks: In Definition 1, one can see that if the function is constant, then it becomes zero, and the kernel does not have any kind of singularity at $t = \tau$.

Definition 2. The Laplace transform (LT) of CF fractional derivative is given as [10]

$$L\left\{{}^{CF}D_t^{\rho}\Psi(\xi,t)\right\}(s) = \frac{B(\rho)sL(\Psi(\xi,t)) - \Psi(\xi,0)}{s+\rho(1-s)}. \tag{4}$$

In general,

$$L\left\{{}^{CF}D_t^{(\rho+n)}\Psi(\xi,t)\right\}(s) = \frac{B(\rho)s^{(n+1)}L\left\{\psi(\xi,t)\right\} - \sum_{k=0}^{n} s^{n-k}\Psi^{k}(\xi,0)}{s+\rho(1-s)}. \tag{5}$$

3. Implementation of the MHATM to Time-Fractional SWEs

In this part, the accuracy and exactness of the MHATM are checked through coupled time-fractional SWEs. The basic theories and detailed steps of MHATM are presented in previous works [20, 21]. Consider the following form of the coupled time-fractional SWEs with CF fractional derivative [17, 18]:

$$\begin{cases} {}^{CF}D_t^{\rho}\Psi(\xi,t) + \Phi(\xi,t)\Psi_{\xi}(\xi,t) + \Psi(\xi,t)\Phi_{\xi}(\xi,t) = 0, \\ {}^{CF}D_t^{\delta}\Phi(\xi,t) + \Phi(\xi,t)\Phi_{\xi}(\xi,t) + \Psi_{\xi}(\xi,t) = 0, \end{cases} \tag{6}$$

subject to primary conditions [17, 18]

$$\Psi(\xi,0) = \frac{1}{9}\left(\xi^2 - 2\xi + 1\right) \quad \text{and} \quad \Phi(\xi,0) = \frac{2}{3}\left(1-\xi\right). \tag{7}$$

The exact solutions for the Eq. (6) when $\rho = 1$ and $\delta = 1$ are given as [17, 18]

$$\Psi(\xi,t) = \frac{(\xi-1)^2}{9(t-1)^2} \quad \text{and} \quad \Phi(\xi,t) = \frac{2(\xi-1)}{3(t-1)}. \tag{8}$$

In accordance with the method, applying LT on both sides of the Eq. (6) and using Definition 2, we get

$$\begin{cases} L[\Psi] - \frac{\Psi(\xi,0)}{s} + \left[\frac{1+\rho(\frac{1}{s}-1)}{B(\rho)}\right] L\left[\Phi(\xi,t)\Psi_{\xi}(\xi,t) + \Psi(\xi,t)\Phi_{\xi}(\xi,t)\right] = 0, \\ L[\Phi] - \frac{\Phi(\xi,0)}{s} + \left[\frac{1+\delta(\frac{1}{s}-1)}{B(\delta)}\right] L\left[\Phi(\xi,t)\Phi_{\xi}(\xi,t) + \Psi_{\xi}(\xi,t)\right] = 0. \end{cases} \tag{9}$$

The nonlinear operators are given as

$$\begin{cases} M_1[\Upsilon_1(\xi,t;q)] = L[\Upsilon_1(\xi,t;q)] - \frac{\Psi(\xi,0)}{s} + \left[\frac{1+\rho(\frac{1}{s}-1)}{B(\rho)}\right] \\ \qquad \times L\left[\Phi(\xi,t)\Psi_{\xi}(\xi,t) + \Psi(\xi,t)\Phi_{\xi}(\xi,t)\right] = 0, \\ M_2[\Upsilon_2(\xi,t;q)] = L[\Upsilon_2(\xi,t;q)] - \frac{\Phi(\xi,0)}{s} + \left[\frac{1+\delta(\frac{1}{s}-1)}{B(\delta)}\right] \\ \qquad \times L\left[\Phi(\xi,t)\Phi_{\xi}(\xi,t) + \Psi_{\xi}(\xi,t)\right] = 0, \end{cases} \tag{10}$$

and consequently, as discussed in [20, 21], we have

$$\begin{cases} \Theta_{1,m-1}[\vec{\Psi}_{m-1}] = L[\vec{\Psi}_{m-1}] - \frac{\psi(\xi,0)}{s} + \left[\frac{1+\delta(\frac{1}{s}-1)}{B(\delta)}\right] \\ \qquad \times L\left[H_{1,m} + H_{2,m}\right] = 0, \\ \Theta_{2,m-1}[\vec{\theta}_{m-1}] = L[\vec{\theta}_{m-1}] - \frac{\theta(\xi,0)}{s} + \left[\frac{1+\nu(\frac{1}{s}-1)}{B(\nu)}\right] \\ \qquad \times L\left[H_{3,m} + (\Psi_{m-1})_{\xi}\right] = 0. \end{cases} \tag{11}$$

Next, a modification has been done by expanding the nonlinear term present in Eq. (11) as a homotopy polynomial and defined as

$$\begin{cases} H_{1,m} = \frac{1}{\Gamma(m+1)} \left[\frac{\partial^m}{\partial q^m} N\left[(q\zeta(\xi,t;q))(q\varsigma(\xi,t;q))_{\xi}\right]\right]_{q=0}, \\ H_{2,m} = \frac{1}{\Gamma(m+1)} \left[\frac{\partial^m}{\partial q^m} N\left[\left((q\varsigma(\xi,t;q))(q\zeta(\xi,t;q))\right)_{\xi}\right]\right]_{q=0}, \\ H_{3,m} = \frac{1}{\Gamma(m+1)} \left[\frac{\partial^m}{\partial q^m} N\left[\left((q\zeta(\xi,t;q))(q\zeta(\xi,t;q))\right)_{\xi}\right]\right]_{q=0}, \end{cases} \tag{12}$$

where

$$\begin{cases} \zeta(\xi,t;q) = \zeta_0 + q\zeta_1 + q^2\zeta_2 + q^3\zeta_3 + \cdots, \\ \varsigma(\xi,t;q) = \varsigma_0 + q\varsigma_1 + q^2\varsigma_2 + q^3\varsigma_3 + \cdots. \end{cases} \tag{13}$$

Hence, we find the mth-order deformation equations as follows:

$$\begin{cases} L[\Psi_m - \chi_m \Psi_{m-1}] = \hbar\, \Theta_{1,m-1}[\vec{\Psi}_{m-1}], \\ L[\Phi_m - \chi_m \Phi_{m-1}] = \hbar\, \Theta_{2,m-1}[\vec{\Phi}_{m-1}], \end{cases} \tag{14}$$

where $\hbar$ is an auxiliary constant known as converging restraint parameter (for more details, see [23, 24]). Next, by taking the inversion of LT in Eq. (14), we get

$$\begin{cases} \Psi_m = \chi_m \Psi_{m-1} + \hbar L^{-1}[\Theta_{1,m-1}(\vec{\Psi}_{m-1})], \\ \Phi_m = \chi_m \Phi_{m-1} + \hbar L^{-1}[\Theta_{2,m-1}(\vec{\Phi}_{m-1})]. \end{cases} \tag{15}$$

With the help of Eqs. (7) and (15) is solved for the different values of $m = 1, 2, 3, \ldots$.
For $m = 1$, we have

$$\begin{cases} \Psi_1 = \frac{-\hbar\ (1-\rho+\rho t)}{B(\delta)} \frac{2}{9} (\xi - 1)^2, \\ \Phi_1 = \frac{-\hbar\ (1-\delta+\delta t)}{B(\delta)} \frac{2}{3} (1 - \xi). \end{cases}$$

For $m = 2$, we have

$$\begin{cases} \Psi_2 = \dfrac{-\hbar(1+\hbar)(1-\rho+\rho t)}{B(\rho)} \dfrac{2}{9} (\xi - 1)^2 + \dfrac{\hbar^2 (1-\rho+\rho t)^2}{B^2(\rho)} \dfrac{4}{9} (\xi - 1)^2 \\ \quad + \dfrac{\hbar^2\ (1-\rho-\delta+\rho+\delta t)}{B(\rho+\delta)} \dfrac{2}{9} (\xi - 1)^2, \\ \Phi_2 = \dfrac{-\hbar(1+\hbar)\ (1-\delta+\delta t)}{B(\delta)} \dfrac{2}{3} (1-\xi) + \dfrac{\hbar^2\ (1-\delta+\delta t)^2}{B^2(\delta)} \dfrac{8}{9} (1-\xi) \\ \quad + \dfrac{\hbar^2\ (1-\rho-\delta+\rho+\delta t)}{B(\rho+\delta)} \dfrac{4}{9} (1-\xi)^2. \end{cases}$$

Therefore, the final solutions of Eq. (6) are

$$\begin{cases} \Psi(\xi,t) = \Psi_0 + \Psi_1 + \Psi_2 + \cdots, \\ \Phi(\xi,t) = \Phi_0 + \Phi_1 + \Phi_2 + \cdots. \end{cases} \tag{16}$$

4. Convergence Analysis of MHATM Solution

The final solution of time-fractional coupled SWEs (Eq. (6)) with CF derivative can be expressed as

$$\begin{cases} \Psi(\xi,t) = \sum_{m=0}^{\infty} \Psi_m(\xi,t), \\ \Phi(\xi,t) = \sum_{m=0}^{\infty} \Phi_m(\xi,t), \end{cases}$$

where

$$\begin{cases} \Psi_m(\xi,t) = (\chi_m + \hbar)\Psi_{m-1} + \hbar(1-\chi_m)\Psi_{m-1} + \hbar L^{-1}\Big[\Big(\frac{1+\rho(\frac{1}{s}-1)}{B(\rho)}\Big) \\ \qquad \times L\Big(\Phi\Psi_\xi + \Psi\Phi_\xi\Big)\Big], \\ \Phi_m(\xi,t) = (\chi_m + \hbar)\Phi_{m-1} + \hbar(1-\chi_m)\Phi_{m-1} + \hbar L^{-1}\Big[\Big(\frac{1+\delta(\frac{1}{s}-1)}{B(\delta)}\Big) \\ \qquad \times L\Big(\Phi\Phi_\xi + \Psi_\xi\Big)\Big]. \end{cases} \tag{17}$$

For simplicity, we make use of the following symbols: $V(\rho) = \frac{1+\rho(\frac{1}{s}-1)}{B(\rho)}$, $V(\delta) = \frac{1+\delta(\frac{1}{s}-1)}{B(\delta)}$, $N_1[\Psi] = \Phi\Psi_\xi$, $N_2[\Psi] = \Psi\Phi_\xi$, $N_3[\Phi] = \Phi\Phi_\xi$, and $M_1[\Phi] = \Psi_\xi$.

Then, Eq. (17) becomes

$$\begin{cases} \Psi_m = (\chi_m + \hbar)\Psi_{m-1} + \hbar(1-\chi_m)\Psi_{m-1} \\ \qquad + \hbar L^{-1}\big[V(\rho)L\big(N_1[\Psi] + N_2[\Psi]\big)\big], \\ \Phi_m = (\chi_m + \hbar)\Phi_{m-1} + \hbar(1-\chi_m)\Phi_{m-1} \\ \qquad + \hbar L^{-1}\big[V(\delta)L\big(N_3[\Phi] + M_1[\Phi]\big)\big], \end{cases} \tag{18}$$

where $N_1[\Psi]$, $N_2[\Psi]$, $N_3[\Phi]$ are the nonlinear operators and $M_1[\Phi]$ is the linear operator. Further, consider that the nonlinear operators $N_1[\Psi]$, $N_2[\Psi]$, and $N_3[\Phi]$ are Lipschitz continuous, i.e., for any two arbitrary functions Ψ_1, Ψ_2 and Φ_1, Φ_2, we find $\gamma_1 > 0, \gamma_2 > 0$, $\gamma_3 > 0$

such that

$$|N_1(\Psi_1) - N_1(\Psi_2)| \leq \gamma_1 |\Psi_1 - \Psi_2|,$$

$$|N_2(\Psi_1) - N_2(\Psi_2)| \leq \gamma_2 |\Psi_1 - \Psi_2|,$$

$$|N_3(\Phi_1) - N_3(\Phi_2)| \leq \gamma_3 |\Phi_1 - \Phi_2|,$$

and the linear operator $M_1[\Phi]$ is a bounded operator in ξ, i.e., it is possible to find number $\alpha_1 > 0$ such that

$$\|M_1 \Phi\| \leq \alpha_1 \|\Phi\|.$$

By using the aforesaid results, we present the following theorems.

Theorem 1 (Uniqueness theorem). The solution acquired by MHATM of the time-fractional SWEs (Eq. (6)) with CF derivative is unique, when $0 < \beta_1, \beta_2 < 1$, where $\beta_1 = (1+\hbar) + \hbar T(\gamma_1 + \gamma_2)$, $\beta_2 = (1+\hbar) + \hbar T(\gamma_3 + \alpha_1)$, and γ_1, γ_2, γ_3, and α_1 are constants.

Proof. If possible, let Ψ, Ψ^* and Φ, Φ^* be the two distinct set of solutions of Eq. (6), then

$$\begin{cases} |\Psi - \Psi^*| = \big|(1+\hbar)(\Psi - \Psi^*) + \hbar L^{-1}\big(V(\rho)L\big(N_1(\Psi - \Psi^*) \\ \qquad\qquad + N_2(\Psi - \Psi^*)\big)\big)\big|, \\ |\Phi - \Phi^*| = \big|(1+\hbar)(\Phi - \Phi^*) + \hbar L^{-1}\big(V(\delta)L\big(N_3(\Phi - \Phi^*) \\ \qquad\qquad + M_1(\Phi - \Phi^*)\big)\big)\big|. \end{cases} \tag{19}$$

Employing convolution theorem to Eq. (19), we get

$$\begin{cases} |\Psi - \Psi^*| = (1+\hbar)\,|\Psi - \Psi^*| \\ \qquad\qquad + \hbar \displaystyle\int_0^t \left[|N_1(\Psi - \Psi^*) + N_2(\Psi - \Psi^*)| \frac{1 - \rho + \rho(t-\tau)}{B(\rho)} \right] d\tau, \\ |\Phi - \Phi^*| = (1+\hbar)\,|\Phi - \Phi^*| \\ \qquad\qquad + \hbar \displaystyle\int_0^t \left[|N_3(\Phi - \Phi^*) + M_1(\Phi - \Phi^*)| \frac{1 - \delta + \delta(t-\tau)}{B(\delta)} \right] d\tau, \end{cases}$$

and so,

$$\begin{cases} |\Psi - \Psi^*| \leq (1+\hbar)|\Psi - \Psi^*| + \hbar \displaystyle\int_0^t \Bigg[|\gamma_1(\Psi - \Psi^*) + \gamma_2(\Psi - \Psi^*)| \\ \qquad \times \dfrac{1-\rho+\rho(t-\tau)}{B(\rho)} \Bigg] d\tau, \\ |\Phi - \Phi^*| \leq (1+\hbar)|\Phi - \Phi^*| + \hbar \displaystyle\int_0^t \Bigg[|\gamma_3(\Phi - \Phi^*) + \alpha_1(\Phi - \Phi^*)| \\ \qquad \times \dfrac{1-\delta+\delta(t-\tau)}{B(\delta)} \Bigg] d\tau. \end{cases}$$

Using the integral mean-value theorem in the above equation, we get

$$\begin{cases} |\Psi - \Psi^*| \leq (1+\hbar)|\Psi - \Psi^*| + \hbar(\gamma_1|\Psi - \Psi^*| + \gamma_2|\Psi - \Psi^*|)T, \\ |\Phi - \Phi^*| \leq (1+\hbar)|\Phi - \Phi^*| + \hbar(\gamma_3|\Phi - \Phi^*| + \alpha_1|\Phi - \Phi^*|)T. \end{cases}$$

Therefore,

$$\begin{cases} |\Psi - \Psi^*| \leq |\Psi - \Psi^*|\beta_1, \\ |\Phi - \Phi^*| \leq |\Phi - \Phi^*|\beta_2, \end{cases}$$

from which we get $(1-\beta_1)|\Psi - \Psi^*| \leq 0$ and $(1-\beta_2)|\Phi - \Phi^*| \leq 0$. Since $0 < \beta_1, \beta_2 < 1$, then $|\Psi - \Psi^*| = 0$ and $|\Phi - \Phi^*| = 0$ implies $\Psi = \Psi^*$ and $\Phi = \Phi^*$. Hence, the statement. □

Theorem 2 (Convergence theorem). Let $(\Xi[J], \|\bullet\|)$ denote a Banach space (BS) of all continuous functions on J and $N : \Xi \to \Xi$ is a nonlinear mapping with

$$\begin{cases} \|N(\Psi_1) - N(\Psi_2)\| \leq \gamma_1 \|\Psi_1 - \Psi_2\| \ \forall\ \Psi_1,\ \Psi_2 \in \Psi, \\ \|N(\Phi_1) - N(\Phi_2)\| \leq \gamma_2 \|\Phi_1 - \Phi_2\| \ \forall\ \Phi_1,\ \Phi_2 \in \Phi. \end{cases}$$

Then, according to Banach's fixed-point theorem, N has a unique fixed point.

Moreover, the sequence originated by the MHATM with an arbitrary selection of $\Psi_0, \Phi_0 \in \Xi$ converges to the fixed point of N, and

$$\begin{cases} \|\Psi_m - \Psi_n\| \leq \dfrac{\gamma_1^n}{1-\gamma_1} \|\Psi_1 - \Psi_0\| \ \forall\ \Psi \in \Xi, \\ \|\Phi_m - \Phi_n\| \leq \dfrac{\gamma_2^n}{1-\gamma_2} \|\Phi_1 - \Phi_0\| \ \forall\ \Phi \in \Xi. \end{cases}$$

Proof. Given that $(\Xi[J], \|\bullet\|)$ BS of all continuous functions on J. The norm is defined in this BS as $\|f(t)\| = \max_{t\in J}|f(t)|$.

To prove the statement, it is sufficient to show that both the sequences $\{\Psi_n\}$ and $\{\Phi_n\}$ generated by MHATM are Cauchy sequences in $(\Xi[J], \|\bullet\|)$. For this,

$$\begin{cases} \|\Psi_m - \Psi_n\| = \max_{t\in J}|\Psi_m - \Psi_n|, \\ \|\Phi_m - \Phi_n\| = \max_{t\in J}|\Phi_m - \Phi_n|. \end{cases}$$

So,

$$\begin{cases} \|\Psi_m - \Psi_n\| = \max_{t\in J}\big|(1+\hbar)(\Psi_{m-1} - \Psi_{n-1}) \\ \qquad\qquad + \hbar L^{-1}\big[V(\rho)L\big(N_1(\Psi_{m-1} - \Psi_{n-1}) \\ \qquad\qquad + N_2(\Psi_{m-1} - \Psi_{n-1})\big)\big]\big|, \\ \|\Phi_m - \Phi_n\| = \max_{t\in J}\big|(1+\hbar)(\Phi_{m-1} - \Phi_{n-1}) \\ \qquad\qquad + \hbar L^{-1}\big[V(\delta)L\big(N_3(\Phi_{m-1} - \Phi_{n-1}) \\ \qquad\qquad + M_1(\Phi_{m-1} - \Phi_{n-1})\big)\big]\big|, \end{cases}$$

and

$$\begin{cases} \|\Psi_m - \Psi_n\| \le \max_{t\in J}\big[(1+\hbar)|\Psi_{m-1} - \Psi_{n-1}| \\ \qquad\qquad + \hbar L^{-1}\big(V(\rho)L\big(N_1|\Psi_{m-1} - \Psi_{n-1}| \\ \qquad\qquad + N_2|\Psi_{m-1} - \Psi_{n-1}|\big)\big)\big], \\ \|\Phi_m - \Phi_n\| \le \max_{t\in J}\big[(1+\hbar)|\Phi_{m-1} - \Phi_{n-1}| \\ \qquad\qquad + \hbar L^{-1}\big(V(\delta)L\big(N_3|\Phi_{m-1} - \Phi_{n-1}| \\ \qquad\qquad + M_1|\Phi_{m-1} - \Phi_{n-1}|\big)\big)\big]. \end{cases} \tag{20}$$

Applying convolution theorem to Eq. (20), we get

$$\begin{cases} |\Psi_m - \Psi_n| = (1+\hbar)\,|\Psi_{m-1} - \Psi_{n-1}| + \hbar \displaystyle\int_0^t \Bigg[|N_1(\Psi_{m-1} - \Psi_{n-1}) \\ \qquad\qquad + N_2(\Psi_{m-1} - \Psi_{n-1})| \dfrac{1-\rho+\rho(t-\tau)}{B(\rho)} \Bigg] d\tau, \\ |\Phi_m - \Phi_n| = (1+\hbar)\,|\Phi_{m-1} - \Phi_{n-1}| + \hbar \displaystyle\int_0^t \Bigg[|N_3(\Phi_{m-1} - \Phi_{n-1}) \\ \qquad\qquad + M_1(\Phi_{m-1} - \Phi_{n-1})| \dfrac{1-\delta+\delta(t-\tau)}{B(\delta)} \Bigg] d\tau, \end{cases}$$

and so,

$$\begin{cases} |\Psi_m - \Psi_n| \leq (1+\hbar)\,|\Psi_{m-1} - \Psi_{n-1}| + \hbar \displaystyle\int_0^t \Bigg[|\gamma_1(\Psi_{m-1} - \Psi_{n-1}) \\ \qquad\qquad + \gamma_2(\Psi_{m-1} - \Psi_{n-1})| \dfrac{1-\rho+\rho(t-\tau)}{B(\rho)} \Bigg] d\tau, \\ |\Phi_m - \Phi_n| \leq (1+\hbar)\,|\Phi_{m-1} - \Phi_{n-1}| + \hbar \displaystyle\int_0^t \Bigg[|\gamma_3(\Phi_{m-1} - \Phi_{n-1}) \\ \qquad\qquad + \alpha_1(\Phi_{m-1} - \Phi_{n-1})| \dfrac{1-\delta+\delta(t-\tau)}{B(\delta)} \Bigg] d\tau. \end{cases}$$

With the help of integral mean-value theorem, the above equation becomes

$$\begin{cases} |\Psi_m - \Psi_n| \leq (1+\hbar)\,|\Psi_{m-1} - \Psi_{n-1}| + \hbar(\gamma_1\,|\Psi_{m-1} - \Psi_{n-1}| \\ \qquad\qquad + \gamma_2\,|\Psi_{m-1} - \Psi_{n-1}|)T, \\ |\Phi_m - \Phi_n| \leq (1+\hbar)\,|\Phi_{m-1} - \Phi_{n-1}| + \hbar(\gamma_3\,|\Phi_{m-1} - \Phi_{n-1}| \\ \qquad\qquad + \alpha_1\,|\Phi_{m-1} - \Phi_{n-1}|)T. \end{cases}$$

Therefore,

$$\begin{cases} |\Psi_m - \Psi_n| \leq |\Psi_{m-1} - \Psi_{n-1}|\,\gamma_1, \\ |\Phi_m - \Phi_n| \leq |\Phi_{m-1} - \Phi_{n-1}|\,\gamma_2. \end{cases}$$

Taking $m = n + 1$, so

$$\begin{cases} \|\Psi_{n+1} - \Psi_n\| \leq \gamma_1 |\Psi_n - \Psi_{n-1}| \leq \gamma_1^2 |\Psi_{n-1} \\ \qquad\qquad - \Psi_{n-2}| \leq \cdots \leq \gamma_1^n |\Psi_1 - \Psi_0|, \\ \|\Phi_{n+1} - \Phi_n\| \leq \gamma_2 |\Phi_n - \Phi_{n-1}| \leq \gamma_2^2 |\Phi_{n-1} \\ \qquad\qquad - \Phi_{n-2}| \leq \cdots \leq \gamma_2^n |\Phi_1 - \Phi_0|. \end{cases}$$

With the aid of triangle inequality, we have

$$\begin{aligned} \|\Psi_m - \Psi_n\| &\leq |\Psi_{n+1} - \Psi_n| + |\Psi_{n+2} - \Psi_{n+1}| + \cdots + |\Psi_m - \Psi_{m-1}| \\ &\leq \left[\gamma_1^n + \gamma_1^{n+1} + \cdots + \gamma_1^{m-1}\right] \|\Psi_1 - \Psi_0\| \\ &\leq \gamma_1^n \left[1 + \gamma_1 + \gamma_1^2 + \cdots + \gamma_1^{m-n-1}\right] \|\Psi_1 - \Psi_0\| \\ &\leq \gamma_1^n \left[\frac{1 - \gamma_1^{m-n-1}}{1 - \gamma_1}\right] \|\Psi_1 - \Psi_0\|. \end{aligned}$$

Similarly,

$$\|\Phi_m - \Phi_n\| \leq \gamma_2^n \left[\frac{1 - \gamma_2^{m-n-1}}{1 - \gamma_2}\right] \|\Phi_1 - \Phi_0\|,$$

i.e., we have

$$\begin{cases} \|\Psi_m - \Psi_n\| \leq \gamma_1^n \left[\frac{1-\gamma_1^{m-n-1}}{1-\gamma_1}\right] \|\Psi_1 - \Psi_0\|, \\ \|\Phi_m - \Phi_n\| \leq \gamma_2^n \left[\frac{1-\gamma_2^{m-n-1}}{1-\gamma_2}\right] \|\Phi_1 - \Phi_0\|. \end{cases}$$

As $0 < \gamma_1, \gamma_2 < 1$; therefore, $(1 - \gamma_1^{m-n}) < 1$ and $(1 - \gamma_2^{m-n}) < 1$. Subsequently, we have

$$\begin{cases} \|\Psi_m - \Psi_n\| \leq \dfrac{\gamma_1^n}{1 - \gamma_1} \|\Psi_1 - \Psi_0\|, \\ \|\Phi_m - \Phi_n\| \leq \dfrac{\gamma_2^n}{1 - \gamma_2} \|\Phi_1 - \Phi_0\|. \end{cases}$$

But $\|\Psi_1 - \Psi_0\| < \infty$ and $\|\Phi_1 - \Phi_0\| < \infty$, so as $m \to \infty$, then $\|\Psi_m - \Psi_n\| \to 0$ and $\|\Phi_m - \Phi_n\| \to 0$. Hence, $\{\Psi_n\}$ and $\{\Phi_n\}$ are both Cauchy sequences in $(\Xi[J], \|\bullet\|)$. Therefore, the sequence converges. □

5. Numerical Results and Discussions

In this section of the chapter, the results obtained for the coupled SWEs by MHATM have been verified through surface graphs and tabulated data. Figures 1 and 2 give the surface graphs of the solutions of the coupled SWEs with the variation of the diverse values of the fractional parameters ρ and δ. These figures show that despite the diverse values of the fractional parameters ρ and δ, their surface graphs are almost the same in nature, which reflects the exactness of the MHATM with CF derivative. Furthermore, the numerical values of the Absolute Error (AE) is shown in Tables 1 and 2. These tables show that the obtained solutions by the proposed method are very much close to the exact ones.

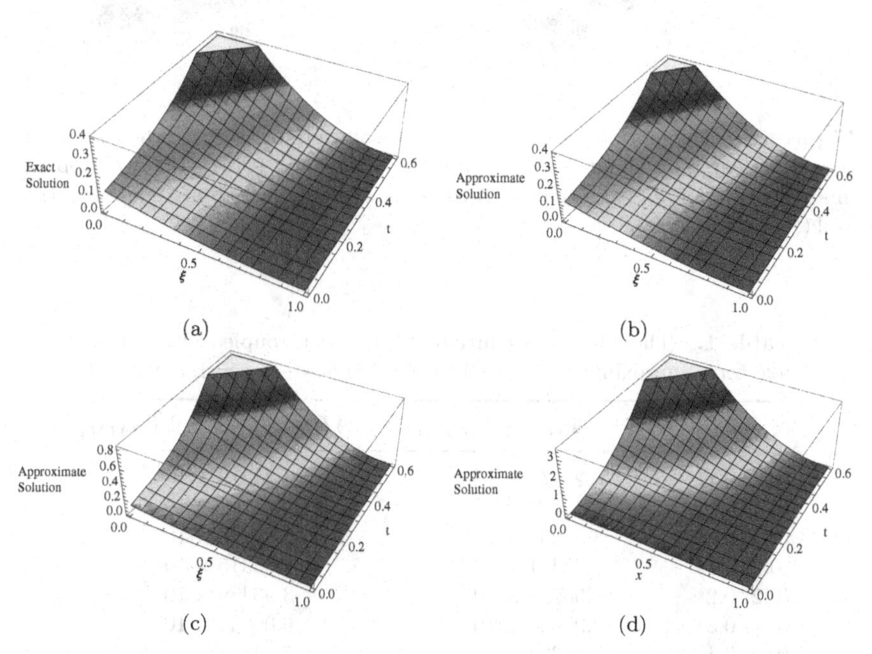

Figure 1. The surfaces show (a) exact solution of $\Psi(\xi, t)$ when $\rho = 1$, (b) numerical approximate solution of $\Psi(\xi, t)$ when $\rho = 1$, (c) numerical approximate solution of $\Psi(\xi, t)$ when $\rho = 0.75$, and (d) numerical approximate solution of $\Psi(\xi, t)$ when $\rho = 0.5$.

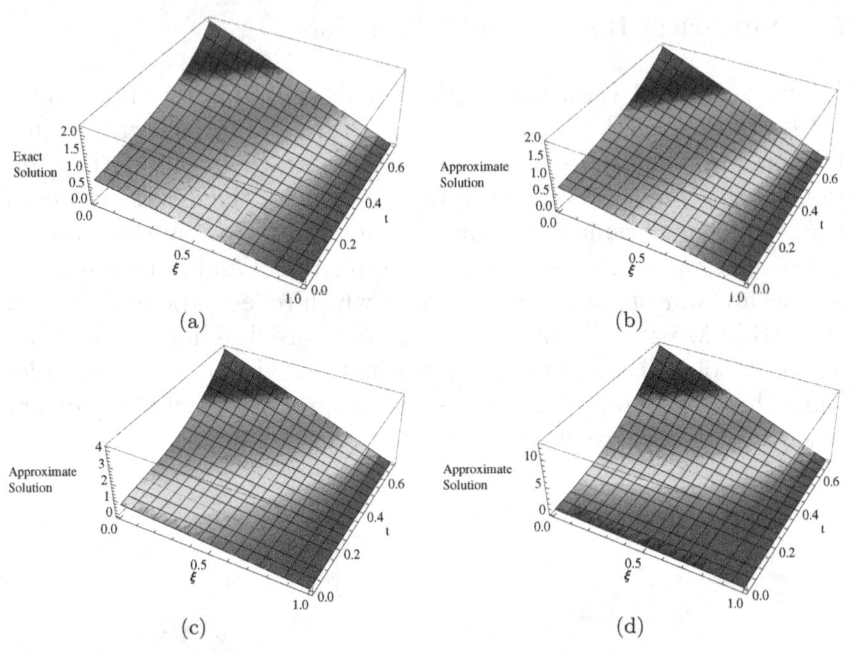

Figure 2. The surfaces show (a) exact solution of $\Phi(\xi, t)$ when $\delta = 1$, (b) numerical approximate solution of $\Phi(\xi, t)$ when $\delta = 1$, (c) numerical approximate solution of $\Phi(\xi, t)$ when $\delta = 0.75$, and (d) numerical approximate solution of $\Phi(\xi, t)$ when $\delta = 0.5$.

Table 1. The AE in the solution of fractional, coupled SWEs using five-term approximation for MHATM when $\hbar = -1$, $\rho = 1$, and $\delta = 1$.

$(\xi,\ t)$	$AE = \left\lvert \Psi_{exact} - \Psi^5_{MHATM} \right\rvert$	$AE = \left\lvert \Phi_{exact} - \Phi^5_{MHATM} \right\rvert$
(0.1, 0.1)	6.21112×10^{-7}	4.98213×10^{-7}
(0.1, 0.2)	4.22476×10^{-5}	3.45290×10^{-5}
(0.1, 0.3)	3.97267×10^{-4}	5.68244×10^{-4}
(0.2, 0.1)	4.86421×10^{-7}	4.85315×10^{-7}
(0.2, 0.2)	3.50985×10^{-5}	3.95439×10^{-5}
(0.2, 0.3)	2.98754×10^{-4}	6.09842×10^{-4}
(0.3, 0.1)	5.86324×10^{-7}	7.34218×10^{-7}
(0.3, 0.2)	2.68923×10^{-5}	5.78531×10^{-5}
(0.3, 0.3)	3.35813×10^{-4}	6.75210×10^{-4}

Table 2. The AE in the solution of fractional, coupled SWEs using five-term approximation for MHATM when $\hbar = -1$, $\rho = 0.75$, and $\delta = 0.75$.

(ξ, t)	$AE = \left|\Psi_{exact} - \Psi^{5}_{MHATM}\right|$	$AE = \left|\Phi_{exact} - \Phi^{5}_{MHATM}\right|$
(0.1, 0.1)	6.34212×10^{-3}	4.32198×10^{-2}
(0.1, 0.2)	3.45210×10^{-2}	6.76213×10^{-2}
(0.1, 0.3)	5.78213×10^{-2}	7.94321×10^{-2}
(0.2, 0.1)	3.67321×10^{-3}	4.56723×10^{-2}
(0.2, 0.2)	2.89213×10^{-2}	6.43216×10^{-2}
(0.2, 0.3)	5.98432×10^{-2}	6.54375×10^{-2}
(0.3, 0.1)	4.59021×10^{-3}	3.85654×10^{-2}
(0.3, 0.2)	3.21982×10^{-2}	2.95318×10^{-2}
(0.3, 0.3)	3.56312×10^{-2}	5.48570×10^{-2}

6. Conclusion

In this chapter, a reliable numerical scheme has been presented to obtained the solution behavior of the time-fractional, coupled SWEs. The nonlinear terms present in Eq. (6) have been enlarged in terms of homotopy polynomial. As a result, we get a faster rate of convergence of the series solutions. The uniqueness and convergence analysis for the solutions of coupled SWEs are presented in Theorems 1 and 2, respectively. Obtained solutions in Eq. (16) are verified through Figures 1 and 2. Also, numerical values in terms of AE of the solutions are given in Tables 1 and 2. From the above discussion, it can be concluded that the MHATM with CF derivative provides the accurateness to solve various nonlinear, coupled fractional problems, such as SWEs.

References

[1] Jena, R. M., Chakraverty, S., and Baleanu, D. (2020). A novel analytical technique for the solution of time-fractional Ivancevic option pricing model. *Physica A: Statistical Mechanics and its Applications*, 124380, 550.

[2] Kumar, A. and Kumar, S. (2018). A modified analytical approach for fractional discrete KdV equations arising in particle vibrations. *Proceedings of the National Academy of Sciences, India Section A: Physical Sciences*, 88, 95–106.

[3] Shin, J., Tang, J. H., and Wu, M. S. (2008). Solution of shallow-water equations using least-squares finite-element method. *Acta Mechanica Sinica*, 24, 523–532.

[4] Szpilk, C. M. and Kolar, R. L. (2003). Numerical analogs to Fourier and dispersion analysis: Development, verification, and application to the shallow water equations. *Advances in Water Resources*, 26(6), 649–662.

[5] Goswami, A., Singh, J., Kumar, D., and Gupta, S. (2019). An efficient analytical technique for fractional partial differential equations occurring in ion acoustic waves in plasma. *Journal of Ocean Engineering and Science*, 4(2), 85–99.

[6] Saghali, S., Javidi, M., and Saei, F. D. (2019). Analytical solution of a fractional differential equation in the theory of viscoelastic fluids. *International Journal of Applied and Computational Mathematics*, 5, 53.

[7] Miller, K. S. and Ross, B. (1993). *An Introduction to Fractional Calculus and Fractional Differential Equations*, John Wiley and Sons: New York.

[8] Podlubny, I. (1999). *Fractional Differential Equations*, Academic Press: New York.

[9] Muslih, S. I. and Agrawal, O. P. (2010). Riesz fractional derivatives and fractional dimensional space. *International Journal of Theoretical Physics*, 49(2), 270–275.

[10] Caputo, M. and Fabrizio, M. (2015). A new definition of fractional derivative without singular kernel. *Progress in Fractional Differentiation and Applications*, 1, 73–85.

[11] Moore, E. J., Sirisubtawee, S., and Koonprasert, S. (2019). A Caputo–Fabrizio fractional differential equation model for HIV/AIDS with treatment compartment. *Advances in Differential Equations*, 200, 1–20.

[12] Atangana, A. and Baleanu, D. (2016). Caputo-Fabrizio derivative applied to groundwater flow within a confined aquifer. *Journal of Engineering Mechanics*, 143, D4016005.

[13] Kumar, A. and Kumar, S. (2016). Residual power series method for fractional Burger types equations. *Journal of Nonlinear Engineering*, 5, 235–244.

[14] Jena, R. M., Chakraverty, S., and Baleanu, D. (2020). Solitary wave solution for a generalized Hirota-Satsuma coupled KdV and MKdV equations: A semi-analytical approach. *Alexandria Engineering*, 59(5), 2877–2889.

[15] Karunakar, P. and Chakraverty, S. (2018). Solution of interval shallow water wave equations using homotopy perturbation method. *Engineering with Computers*, 34(4), 1610–1624.

[16] Karunakar, P. and Chakraverty, S. (2017). Comparison of solutions of linear and non-linear shallow water wave equations using homotopy perturbation method. *International Journal of Numerical Methods for Heat & Fluid Flow*, 27(9), 2015–2029.

[17] Kumar, S., Kumar, A., Odibat, Z., Aldhaifallah, M., and Nisar, K. S. (2020). A comparison study of two modified analytical approach for the solution of nonlinear fractional shallow water equations in fluid flow. *AIMS Mathematics*, 5(4), 3035–3055.

[18] Kumar, S. (2013). A numerical study for solution of time fractional nonlinear shallow-water equation in oceans. *Zeitschrift für Naturforschung*, 68, 1–7.

[19] Odibat, Z. and Bataineh, A. S. (2015). An adaptation of homotopy analysis method for reliable treatment of strongly nonlinear problems: Construction of homotopy polynomials. *Mathematical Methods in the Applied Sciences*, 38, 991–1000.

[20] Kumar, S., Kumar, A., and Baleanu, D. (2016). Two analytical methods for time-fractional nonlinear coupled Boussinesq-Burgers equations arise in propagation of shallow water waves. *Nonlinear Dynamics*, 85, 699–715.

[21] Kumar, S., Kumar, A., and Argyros, I. K. (2017). A new analysis for the Keller-Segel model of fractional order. *Numerical Algorithms*, 75, 213–228.

[22] Veeresha, P., Prakasha, D. G., and Kumar, D. (2020). An efficient technique for nonlinear time-fractional Klein–Fock–Gordon equation. *Applied Mathematics and Computation*, 364, 124637.

[23] Liao, S. J. (1992). The proposed homotopy analysis technique for the solution of nonlinear problems. Ph.D Thesis, Shanghai Jiao Tong University.

[24] Liao, S. J. (1997). Homotopy analysis method: A new analytical technique for nonlinear problems. *Communications in Nonlinear Science and Numerical Simulation*, 2, 95–100.

https://doi.org/10.1142/9789811245367_0013

Chapter 13

Solution of Fractional Wave Equation by Homotopy Perturbation Method

Shweta Dubey* and S. Chakraverty

Department of Mathematics, National Institute of Technology, Rourkela, Odisha, India

**shwetadubey9452@gmail.com*

Abstract

The general fractional wave equations have been used for modeling different processes in complex or viscoelastic media, disordered material, etc. Two-dimensional fractional wave equation with boundary conditions has been already solved by homotopy perturbation method in other paper. In this chapter, we have solved for both one-dimensional and two-dimensional fractional wave equations with boundary conditions using homotopy perturbation method. The fractional wave equation contains time-fractional derivative of order α, where $1 < \alpha < 2$, and the fractional derivative is described in Caputo sense. This method gives an approximate solution in the form of a convergent series, where components of the series can be easily computed. The results obtained here shows that the introduced homotopy perturbation method is efficient and straightforward to implement. Two numerical examples are illustrated for different values of α to confirm the theoretical results.

Keywords: Caputo sense, fractional wave equation, homotopy perturbation method, Riemann–Liouville fractional integral

1. Introduction

In recent years, it has been found that fractional-order derivatives are very essential for many physical phenomena, such as rheology, damping law, and diffusion process. Several analytical and numerical methods have been proposed to solve fractional-differential equations, such as finite-difference method (FDM) [1], Adomian decomposition method (ADM) [5, 11], Laplace transformation method (LTM) [10, 12], Sumudu transformation method [6], shifted Legendre polynomial-based Galerkin and collocation methods [8], homotopy perturbation method (HPM) [2, 9], etc. The HPM is a very efficient way to find the approximate analytical solutions of fractional-differential equations. HPM was first proposed by the Chinese mathematician Ji-Haun He [3, 4]. The main idea of this method is to introduce a homotopy parameter, say p, which takes values from 0 to 1. When $p = 0$, the differential equation converts to a simplified form which admits a simple solution. As p increases from 0 to 1, the solution is close to the desired one. Eventually, at $p = 1$, the differential equation admits original form of equation and admits the desired solution.

The wave equations are very influential in various fields of science and engineering. Many researchers worked to model wave equations by substituting the fractional derivatives in place of ordinary derivatives. The time-fractional wave equations (TFWEs) are also solved by fractional reduced differential-transformation method [7].

Here, we will describe one- and two-dimensional TFWEs with boundary conditions with the aid of HPM.

The problem based on two-dimensional TFWE with given boundary conditions has been solved by Zhang *et al.* [14], which we have used here for one-dimensional TFWE with different boundary conditions. In that paper, the graphical presentation of the problem has been illustrated for four different values of α (1.20, 1.50, 1.80, and 2), while we have discussed the problem for two different values of α (1.5 and 2) graphically. In this chapter, we have also solved a problem on two-dimensional TFWE with initial condition. Solution to the considered problem has been illustrated graphically for $\alpha = 1.5$ and 2.

The one-dimensional TFWE with boundary conditions are given as

$$\begin{cases} D_t^{\alpha}u(x,t) = u_{xx}(x,t) + f(x,t), & 1 < \alpha \leq 2 \\ u(x,0) = g_1(x), u(x,b) = g_2(x), & x > 0, 0 < t < b \end{cases},$$

and the two-dimensional TFWE with boundary conditions [14] are given as

$$\begin{cases} D_t^{\alpha}u(x,y,t) = u_{xx}(x,y,t) + u_{yy}(x,y,t) + f(x,y,t), \\ \quad 1 < \alpha \leq 2, x, y > 0, 0 < t < a \\ u(x,y,0) = g_1(x,y), u(x,y,a) = g_2(x,y) \end{cases}$$

The main aim of this paper is to apply homotopy perturbation method [3, 4] for solving one-dimensional and two-dimensional TFWEs [14].

The rest of the chapter is structured as follows. In Section 2, preliminaries are given. In Section 3, description of HPM for one-dimensional TFWE is given. In Section 4, an example is solved using HPM. Section 5 consists of the description of HPM for two-dimensional TFWE and a numerical example is given in Section 6. Finally, the conclusion is given in Section 7.

2. Preliminaries

In this section, some basic definitions and properties of fractional-calculus theory [10, 12, 13] are given, which would be further used in the subsequent sections.

Definition 2(a) ([14]). The Riemann–Liouville fractional integral of order $\alpha \geq 0$ of a function $u(t)$ is defined as

$$\begin{cases} J^{\alpha}u(t) = \dfrac{1}{\Gamma\alpha}\displaystyle\int_0^t (t-\tau)^{\alpha}u(\tau)d\tau, & \alpha > 0, \tau > 0 \\ J^0 u(t) = u(t) \end{cases}. \tag{1}$$

For α, $\beta \geq 0$ and $\gamma \geq -1$, we mention some properties of the operator J^{α} here, which are given as follows:

1. $J^\alpha J^\beta u(t) = J^{\alpha+\beta} u(t)$;
2. $J^\alpha J^\beta u(t) = J^\beta J^\alpha u(t)$;
3. $J^\alpha t^\gamma = \frac{\Gamma(\gamma+1)}{\Gamma(\alpha+\gamma+1)} t^{\alpha+\gamma}$.

Definition 2(b) ([14]). The fractional derivative of function $u(t)$ in Caputo sense is defined as follows:

$$D_t^\alpha u(t) = \begin{cases} J^{\alpha-n} \dfrac{d^n}{dt^n} u(t) = \dfrac{1}{\Gamma(n-\alpha)} \\ \qquad \times \displaystyle\int_0^t (t-\tau)^{n-\alpha-1} u^n(\tau) d\tau, \quad n-1 < \alpha < n \\ \dfrac{d^n}{dt^n} u(t), \quad \alpha = n. \end{cases} \tag{2}$$

Lemma 2(c) ([14]). *If $n < \alpha \leq 1$, $n \in N$, then for the function $u(t)$, the following two properties hold:*

$$D_t^\alpha [J^\alpha u(t)] = u(t); 2. \quad J^\alpha [D_t^\alpha u(t)] = u(t) - \sum_{k=0}^{n-1} u^k(0) \frac{t^k}{k!}. \tag{3}$$

3. Homotopy Perturbation Method for One-Dimensional TFWE

In this section, HPM has been described by considering the following one-dimensional TFWE [14]:

$$D_t^\alpha u(x,t) = u_{xx}(x,t) + f(x,t), \quad 1 < \alpha \leq 2, \tag{4}$$

with the boundary conditions

$$u(x,0) = g_1(x), \quad u(x,b) = g_2(x), \quad x > 0, \quad 0 < t < b, \tag{5}$$

where b is a positive constant, D_t^α denotes the Caputo fractional derivative with respect to time, and $f(x,t)$, $g_1(x)$, and $g_2(x)$ are given functions, while $u(x,t)$ is an unknown function such that $u_{xx}(x,t) = \frac{d^2 u(x,t)}{dx^2}$ denotes the second-order derivative of $u(x,t)$ with respect to space variable x.

Equation (4) can be rewritten as

$$D_t^{\alpha} u(x,t) - u_{xx}(x,t) - f(x,t) = 0. \tag{6}$$

Now, constructing the homotopy which satisfies the relation

$$D_t^{\alpha} u(x,t) + p(-u_{xx}(x,t)) - f(x,t) = 0, \tag{7}$$

where $p \in [0,1]$ is the embedding parameter.

If $p = 1$, we get the solution of Eq. (6).

Let the solution of Eq. (7) be given in the form of an infinite-series expansion as follows:

$$u(x,t) = u_0(x,t) + pu_1(x,t) + p^2 u_2(x,t) + \cdots . \tag{8}$$

Substituting Eq. (8) in Eq. (7), we get

$$\begin{aligned} &D_t^{\alpha}(u_0(x,t) + pu_1(x,t) + p^2 u_2(x,t) + \cdots) \\ &\quad = p[D_x^2(u_0(x,t) + pu_1(x,t) + p^2 u_2(x,t) + \cdots)] + f(x,t). \end{aligned} \tag{9}$$

Equating the coefficients of identical power of p from both sides of Eq. (9), we obtain

$$\left.\begin{aligned} &p^0 : D_t^{\alpha} u_0(x,t) = f(x,t) \\ &p^1 : D_t^{\alpha} u_1(x,t) = D_x^2 u_0(x,t) = u_{0xx}(x,t) \\ &p^2 : D_t^{\alpha} u_2(x,t) = D_x^2 u_1(x,t) = u_{1xx}(x,t) \\ &p^3 : D_t^{\alpha} u_3(x,t) = D_x^2 u_2(x,t) = u_{2xx}(x,t) \\ &\vdots \end{aligned}\right\} \tag{10}$$

and so on.

For $p = 1$, we get from Eq. (8), $u(x,t) = \sum_{m=0}^{\infty} u_m(x,t)$.

Applying the inverse operator J^α on both sides of Eq. (10) and using Eq. (3), we get

$$\left.\begin{aligned} u_0(x,t) &= u(x,0) + tu_t(x,0) + J^\alpha[f(x,t)] \\ u_1(x,t) &= J^\alpha[u_{0xx}(x,t)] \\ u_2(x,t) &= J^\alpha[u_{1xx}(x,t)] \\ u_3(x,t) &= J^\alpha[u_{2xx}(x,t)] \\ &\vdots \\ u_{m+1}(x,t) &= J^\alpha[u_{mxx}(x,t)] \\ &\vdots \end{aligned}\right\} \tag{11}$$

and so on.

Substituting Eq. (11) in Eq. (8) for $p = 1$, we obtain

$$\sum_{m=0}^{\infty} u_m(x,t) = u(x,0) + tu_t(x,0) + J^\alpha\left[\sum_{m=0}^{\infty}(u_{mxx}(x,t)) + f(x,t)\right]. \tag{12}$$

Equation (12) can also be rewritten as

$$\sum_{m=0}^{\infty} u_m(x,t) = g_1(x,t) + tu_t(x,0) + J^\alpha[D_x^2(u(x,t)) + f(x,t)], \tag{13}$$

which is the general form of solution for one-dimensional TFWE.

4. Numerical Implementation

Here, we solve a numerical problem of one-dimensional TFWE with boundary conditions using HPM.

Example 1. Consider the following one-dimensional TFWE with boundary conditions:

$$\begin{cases} D_t^\alpha u(x,t) = u_{xx}(x,t) + f(x,t), & x > 0, 0 < t < 1, \\ u(x,0) = x^4, \quad u(x,1) = x^4 + 1 \end{cases}, \tag{14}$$

where $f(x,t) = 12\,t^2 - 12\,x^2$.

Solution: For $\alpha = 2$, the corresponding integer-order problem has the exact solution as $u(x,t) = x^4 + t^4$.

For solving the above problem by HPM, construction of homotopy is required, which satisfies the relation as given in Section 3; therefore, we have

$$\left.\begin{array}{l} D_t^\alpha u(x,t) - 12t^2 - 12x^2 - p[u_{xx}(x,t)] = 0, \quad p \in [0,\,1] \\ u(x,0) = x^4, \quad u(x,1) = x^4 + 1 \end{array}\right\}. \tag{15}$$

Let the solution of Eq. (15) be given as follows:

$$u = u_0 + pu_1 + p^2u_2 + p^3u_3 + \cdots. \tag{16}$$

Substituting Eq. (16) in Eq. (15) and equating the coefficients of like powers of p from both sides, we have the following sets of equations:

$$\left.\begin{array}{l} p^0 : D_t^\alpha u_0(x,t) = 12t^2 - 12x^2 \\ p^1 : D_t^\alpha u_1(x,t) = D_x^2 u_0(x,t) = u_{0xx}(x,t) \\ p^2 : D_t^\alpha u_2(x,t) = D_x^2 u_1(x,t) = u_{1xx}(x,t) \\ p^3 : D_t^\alpha u_3(x,t) = D_x^2 u_2(x,t) = u_{2xx}(x,t) \\ \vdots \end{array}\right\}. \tag{17}$$

Operating J^α on both sides, Eq. (17) gives the following result:

$$u_0(x,t) = x^4 + \frac{12\Gamma(3)t^{\alpha+2}}{\Gamma(\alpha+3)} - \frac{12x^2t^{\alpha+2}}{\Gamma(\alpha+1)}$$

$$u_1(x,t) = \frac{12x^2t^{\alpha+2}}{\Gamma(\alpha+1)} - \frac{24\Gamma(\alpha+1)t^{2\alpha}}{\Gamma(\alpha+2)}$$

$$u_2(x,t) = \frac{24\Gamma(\alpha+1)t^{2\alpha}}{\Gamma(\alpha+2)}$$

$$u_3(x,t) = 0$$

$$u_4(x,t) = 0$$

$$\vdots$$

The rest of components will be $u_5 = u_6 = u_7 = \cdots = 0$.

When $p = 1$, the solution is obtained as

$$\begin{aligned} u &= u_0 + u_1 + u_2 + u_3 + \cdots \\ &= u_0 + u_1 + u_2 \\ &= x^4 + \frac{24t^{\alpha+2}}{\Gamma(\alpha+3)} \end{aligned}$$

For integer-order solution taking $\alpha = 2$, we have $u(x,t) = x^4 + t^4$, which is identical to the exact solution.

For fractional order $\alpha = \frac{3}{2}$, the solution may be written as

$$\begin{aligned} u &= x^4 + \frac{24t^{7/2}}{\Gamma(9/2)} \\ &= x^4 + \frac{128t^{7/2}}{35\sqrt{\Pi}} \end{aligned}$$

The graphical solution of the considered one-dimensional TFWE with boundary conditions has been illustrated for the particular values of $\alpha = 2$ and $\alpha = \frac{3}{2}$ in Figures 1 and 2, respectively.

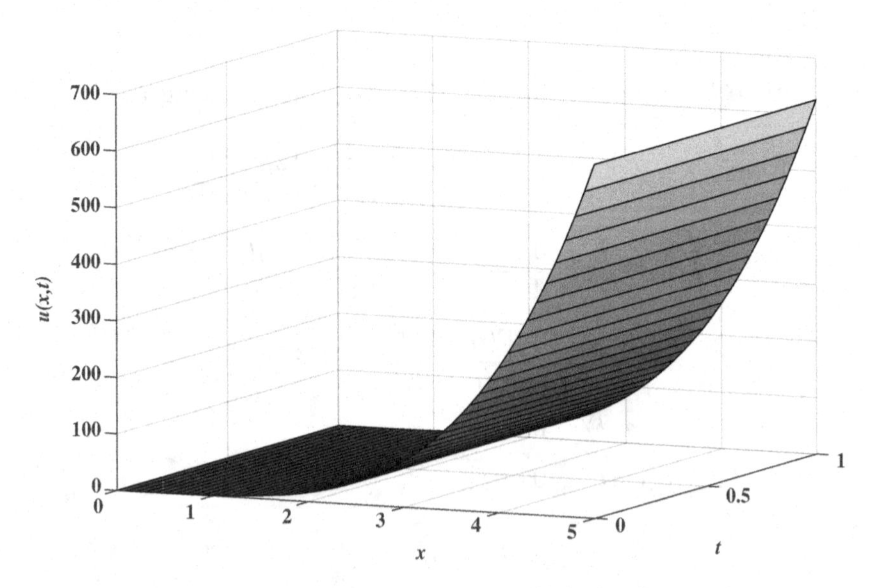

Figure 1. Graphical presentation of solution for $\alpha = 2$.

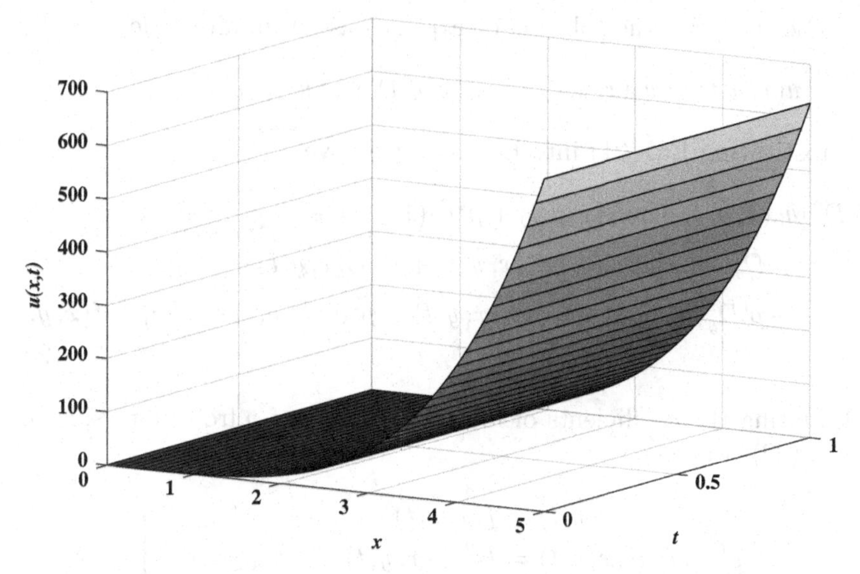

Figure 2. Graphical presentation of solution for $\alpha = \frac{3}{2}$.

5. Homotopy Perturbation Method for Two-Dimensional TFWE

Further, HPM has been applied to solve two-dimensional TFWE with boundary conditions [14].

Consider the following two-dimensional TFWE [14]:

$$D_t^{\alpha} u(x,y,t) = u_{xx}(x,y,t) + u_{yy}(x,y,t) + f(x,y,t),$$
$$1 < \alpha \leq 2, \quad x,y > 0, \quad 0 < t < a, \tag{18}$$

with boundary conditions

$$u(x,y,0) = g_1(x,y), \quad u(x,y,a) = g_2(x,y). \tag{19}$$

First, we construct the homotopy which satisfies the relation

$$D_t^{\alpha} u(x,y,t) = p[u_{xx}(x,y,t) + u_{yy}(x,y,t)] + f(x,y,t), \tag{20}$$

where $p \in [0,1]$.

Let us write the solution of Eq. (20) as an infinite series:

$$u(x,y,t) = u_0(x,y,t) + pu_1(x,y,t) + p^2u_2(x,y,t) + \cdots. \tag{21}$$

Substituting Eq. (21) into Eq. (20), we have

$$\begin{aligned} &D_t^\alpha[u_0(x,y,t) + pu_1(x,y,t) + p^2u_2(x,y,t) + \cdots] \\ &\quad = p[D_x^2(u_0(x,y,t) + pu_1(x,y,t) + p^2u_2(x,y,t) + \cdots)] \\ &\qquad + p[D_y^2(u_0(x,y,t) + pu_1(x,y,t) + p^2u_2(x,y,t) + \cdots)] + f(x,y,t) \end{aligned} \tag{22}$$

Equating the coefficients of identical powers of p from both sides of Eq. (22), we get

$$\left.\begin{aligned} p^0 &: D_t^\alpha u_0(x,y,t) = f(x,y,t) \\ p^1 &: D_t^\alpha u_1(x,y,t) = D_x^2 u_0(x,y,t) + D_y^2 u_0(x,y,t) \\ &\qquad\qquad\qquad = u_{0xx}(x,y,t) + u_{0yy}(x,y,t) \\ p^2 &: D_t^\alpha u_2(x,y,t) = D_x^2 u_1(x,y,t) + D_y^2 u_1(x,y,t) \\ &\qquad\qquad\qquad = u_{1xx}(x,y,t) + u_{1yy}(x,y,t) \\ p^3 &: D_t^\alpha u_3(x,y,t) = D_x^2 u_2(x,y,t) + D_y^2 u_2(x,y,t) \\ &\qquad\qquad\qquad = u_{2xx}(x,y,t) + u_{2yy}(x,y,t) \\ &\vdots \end{aligned}\right\} \tag{23}$$

and so on.

Applying the inverse operator J^α on both sides of Eq. (23), we have

$$\left.\begin{aligned} u_0(x,y,t) &= u(x,y,0) + tu_t(x,y,0) + J^\alpha[f(x,y,t)] \\ u_1(x,y,t) &= J^\alpha[u_{0xx}(x,y,t) + u_{0yy}(x,y,t)] \\ u_2(x,y,t) &= J^\alpha[u_{1xx}(x,y,t) + u_{1yy}(x,y,t)] \\ u_3(x,y,t) &= J^\alpha[u_{2xx}(x,y,t) + u_{2yy}(x,y,t)] \\ &\vdots \\ u_{m+1}(x,y,t) &= J^\alpha[u_{mxx}(x,y,t) + u_{myy}(x,y,t)] \\ &\vdots \end{aligned}\right\} \tag{24}$$

and so on.

Substituting Eq. (24) into Eq. (21) for $p = 1$, we obtain

$$\sum_{m=0}^{\infty} u_m(x,y,t) = u(x,y,0) + t\,u_t(x,y,0) + J^{\alpha}\left[D_x^2 \sum_{m=0}^{\infty}(u_m(x,y,t)) + D_y^2 \sum_{m=0}^{\infty}(u_m(x,y,t)) + f(x,y,t)\right]. \tag{25}$$

Equation (25) can also be rewritten as,

$$\sum_{m=0}^{\infty} u_m(x,y,t) = g_1(x,y,t) + tu_t(x,y,0) + J^{\alpha}[D_x^2(u(x,y,t)) + D_y^2(u(x,y,t)) + f(x,y,t)], \tag{26}$$

which is the general form of solution for two-dimensional TFWE.

6. Numerical Implementation

An example of two-dimensional TFWE with initial condition has been solved here using HPM.

Example 2. Consider the following two-dimensional TFWE with initial condition:

$$\begin{aligned} D_t^{\alpha}u(x,y,t) &= (-1)[u_{xx}(x,y,t) + u_{yy}(x,y,t)] \\ &\quad \times 1 < \alpha \leq 2, \quad x, y > 0, \quad 0 < t < a \\ u(x,y,0) &= -(\cos x + \cos y) \end{aligned} \tag{27}$$

Solution: The exact solution for integer order $\alpha = 2$ of Eq. (27) is given as

$$u(x,y,t) = -(\cos x + \cos y)e^{-t}.$$

To apply HPM, we construct the homotopy as

$$\begin{aligned} D_t^{\alpha}u(x,y,t) &= p(-1)[u_{xx}(x,y,t) + u_{yy}(x,y,t)], \quad p \in [0,1] \\ u(x,y,0) &= -(\cos x + \cos y) \end{aligned} \tag{28}$$

Let us write the solution of Eq. (28) in the form of an infinite series as follows:

$$u(x,y,t) = u_0(x,y,t) + pu_1(x,y,t) + p^2u_2(x,y,t) + \cdots. \tag{29}$$

Putting Eq. (29) into Eq. (28) and equating the coefficients of like powers of p, we have

$$\left.\begin{aligned} p^0 &: D_t^{\alpha} u_0(x,y,t) = 0, \quad u_0(x,y,0) = -(\cos x + \cos y) \\ p^1 &: D_t^{\alpha} u_1(x,y,t) = (-1)[u_{0xx}(x,y,t) + u_{0yy}(x,y,t)], \\ &\qquad u_1(x,y,0) = 0 \\ p^2 &: D_t^{\alpha} u_2(x,y,t) = (-1)[u_{1xx}(x,y,t) + u_{1yy}(x,y,t)], \\ &\qquad u_2(x,y,0) = 0 \\ p^3 &: D_t^{\alpha} u_3(x,y,t) = (-1)[u_{2\,xx}(x,y,t) + u_{2yy}(x,y,t)], \\ &\qquad u_3(x,y,0) = 0 \\ &\vdots \end{aligned}\right\} \tag{30}$$

and so on.

Operating J^{α}, Eq. (30) gives

$$\left.\begin{aligned} u_0(x,y,t) &= [-(\cos x + \cos y)](1-t) \\ u_1(x,y,t) &= [-(\cos x + \cos y)]\left(\frac{t^{\alpha}}{\Gamma(\alpha+1)} - \frac{t^{\alpha+1}}{\Gamma(\alpha+2)}\right) \\ u_2(x,y,t) &= [-(\cos x + \cos y)]\left(\frac{t^{2\alpha}}{\Gamma(2\alpha+1)} - \frac{t^{2\alpha+1}}{\Gamma(2\alpha+2)}\right) \\ &\vdots \end{aligned}\right. \tag{31}$$

and so on.

After substituting the values of Eq. (31) into the infinite series solution in Eq. (29) for $p = 1$, we get

$$\begin{aligned} u(x,y,t) = [-(\cos x + \cos y)] \times \Bigg(1 - t + \frac{t^{\alpha}}{\Gamma(\alpha+1)} - \frac{t^{\alpha+1}}{\Gamma(\alpha+2)} \\ + \frac{t^{2\alpha}}{\Gamma(2\alpha+1)} - \frac{t^{2\alpha+1}}{\Gamma(2\alpha+2)} + \cdots\Bigg) \end{aligned} \tag{32}$$

For $\alpha = 2$, the infinite-series solution in Eq. (32) will be of the form

$$u(x,y,t) = [-(\cos x + \cos y)] \times \left(1 - t + \frac{t^2}{2!} - \frac{t^3}{3!} + \frac{t^4}{4!} - \frac{t^5}{5!} + \cdots\right)$$

$$u(x,y,t) = [-(\cos x + \cos y)]e^{-t},$$

which is the exact solution.

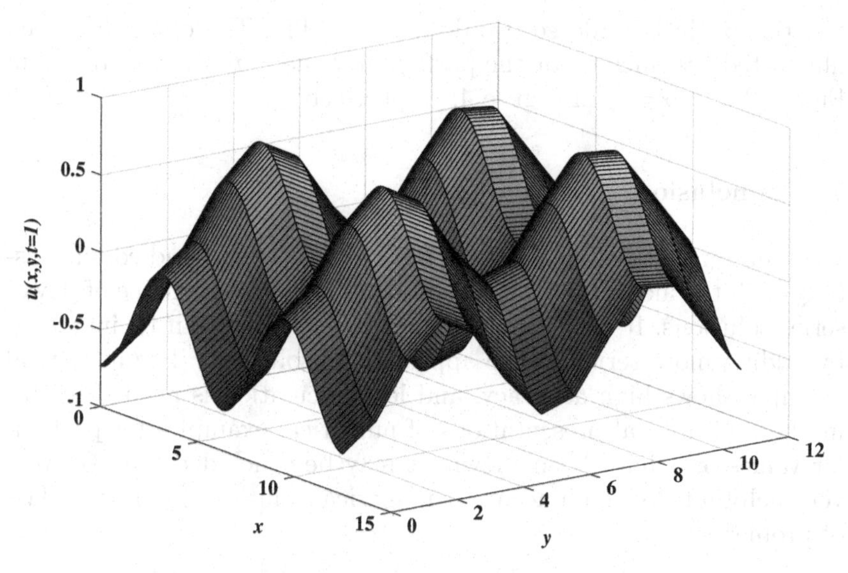

Figure 3. Graphical presentation of solution for $\alpha = 2$.

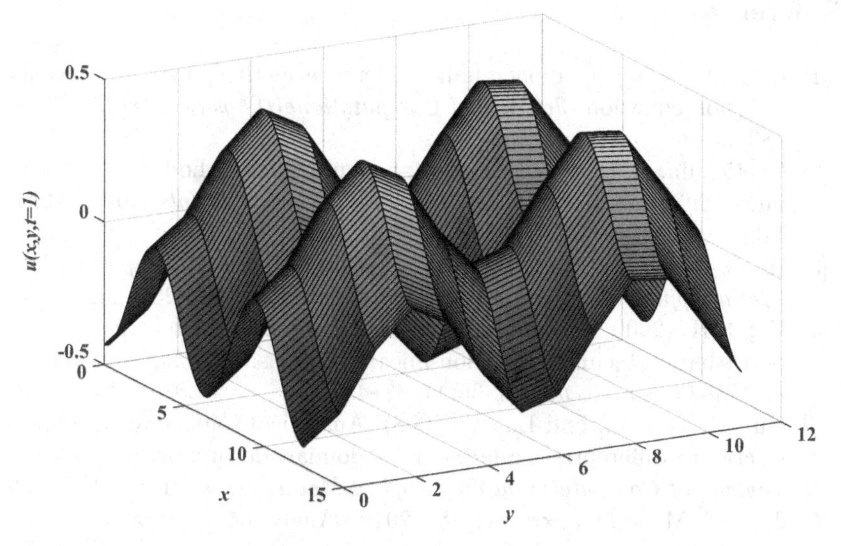

Figure 4. Graphical presentation of solution for $\alpha = \frac{3}{2}$.

For $\alpha = \frac{3}{2}$, the infinite-series solution in Eq. (32) will take the form as

$$u(x, y, t) = [-(\cos x + \cos y)] \times \left(1 - t + \frac{8t^{3/2}}{15\sqrt{\Pi}} - \frac{16t^{5/2}}{105\sqrt{\Pi}} + \cdots\right).$$

Solution of the considered two-dimensional TFWE problem has been illustrated graphically for the particular value of $t = 1$ with $\alpha = 2$ in Figure 3 and $\alpha = \frac{3}{2}$ in Figure 4, respectively.

7. Conclusion

The one- and two-dimensional TFWEs have been handled successfully with the aid of HPM. HPM provides rapid convergence of power series solutions. It may be noted that the accuracy can be increased by adding more terms in the approximate solutions. The numerical example shows high accuracy, and less calculation is needed in this method. Graphical presentations of numerical examples for particular values of α have been shown. It may be concluded that HPM is very helpful to get high-accuracy numerical solutions for a wide class of problems.

References

[1] Cui, M. (2009). Compact finite difference method for the fractional diffusion equation. *Journal of Computational Physics*, 228(20), 7792–7804.
[2] Cveticanin, L. (2006). Homotopy-perturbation method for pure non-linear differential equation. *Chaos, Solitons & Fractals*, 30(5), 1221–1230.
[3] He, J.-H. (1999). Homotopy perturbation technique. *Computer Methods in Applied Mechanics and Engineering*, 178(3–4), 257–262.
[4] He, J.-H. (2000). A coupling method of a homotopy technique and a perturbation technique for non-linear problems. *International Journal of Non-Linear Mechanics*, 35(1), 37–43.
[5] Hu, Y., Luo, Y., and Lu, Z. (2008). Analytical solution of the linear fractional differential equation by Adomian decomposition method. *Journal of Computational and Applied Mathematics*, 215(1), 220–229.
[6] Jena, R. M. and Chakraverty, S. (2019). Analytical solution of Bagley–Torvik equations using Sumudu transformation method. *SN Applied Sciences*, 1(3), 1–6.
[7] Jena, R. M., Chakraverty, S., and Baleanu, D. (2019). On new solutions of time-fractional wave equations arising in shallow water wave propagation. *Mathematics*, 7(8), 722.
[8] Jena, R. M., Chakraverty, S., Edeki, S. O., and Ofuyatan, O. M. (2020). Shifted Legendre polynomial based galerkin and collocation methods

for solving fractional order delay differential equations. *Journal of Theoretical and Applied Information Technology*, 98(4), 535–547.

[9] Alam Khan, N., Ara, A., Anwer Ali, S., and Mahmood, A. (2009). Analytical study of Navier–Stokes equation with fractional orders using he's homotopy perturbation and variational iteration methods. *International Journal of Nonlinear Sciences and Numerical Simulation*, 10(9), 1127–1134.

[10] Miller, K. S. and Ross, B. (1993). *An Introduction to the Fractional Calculus and Fractional Differential Equations*, Wiley, New York.

[11] Momani, S. and Odibat, Z. (2006). Analytical solution of a time-fractional Navier–Stokes equation by Adomian decomposition method. *Applied Mathematics and Computation*, 177(2), 488–494.

[12] Podlubny, I. (1997). The Laplace transform method for linear differential equations of the fractional order. arXiv preprint funct-an/9710005.

[13] Podlubny, I. (1998). *Fractional Differential Equations: An Introduction to Fractional Derivatives, Fractional Differential Equations, to Methods of their Solution and Some of their Applications*, Elsevier, San Diego.

[14] Zhang, X., Zhao, J., Liu, J., and Tang, B. (2014). Homotopy perturbation method for two-dimensional time-fractional wave equation. *Applied Mathematical Modelling*, 38(23), 5545–5552.

https://doi.org/10.1142/9789811245367_bmatter

Index

N

O

P

Q

R

S

T